Telecourse Student Guide
for

Cycles of Life
EXPLORING BIOLOGY

Second Edition

Gerald L. Kellogg

for

**Coast Community College District
Costa Mesa, California**

Brooks/Cole
Thomson Learning™

Pacific Grove • Albany • Belmont • Boston • Cincinnati • Johannesburg • London • Madrid • Melbourne
Mexico City • New York • Scottsdale • Singapore • Tokyo • Toronto

Coast Community College District

William M. Vega, Chancellor

Leslie Purdy, President, Coastline Community College

Peter Vander Haeghen, Administrative Dean, Office of Instructional Systems Development

Laurie R. Melby, Director of Production

Harry Ratner, Producer

Wendy Sacket, Senior Publications Assistant

Thien Vu, Publications Assistant

COPYRIGHT © 2000 by Coast Community College District.
Distributed by Coast Telecourses, Coastline Community College, 11460 Warner Avenue, Fountain Valley, CA 92708. Telephone 800-547-4748. FAX 714-241-6286.

Published by Brooks/Cole
A division of Thomson Learning
The Thomson Learning logo is a trademark used herein under license.

For more information, contact:
BROOKS/COLE
511 Forest Lodge Road
Pacific Grove, CA 93950 USA
www.brookscole.com

All rights reserved. No part of this work covered by the copyright hereon may be reproduced, transcribed or used in any form or by any means—graphic, electronic, or mechanical, including photocopying, recording, taping, Web distribution, or information storage and/or retrieval systems—without the prior written permission of the publisher.

Printed in the United States of America
10   9   8   7   6   5   4   3   2

ISBN 0-534-37272-4

Telecourse Student Guide

for

Cycles of Life
EXPLORING BIOLOGY

Second Edition

Preface

To the Student

Welcome to the telecourse *Cycles of Life: EXPLORING BIOLOGY*, an introductory course in biology. Whether you are planning a career in science, taking the course as part of an academic course for college credit, or taking it just for pleasure, we believe that you will find this telecourse interesting, entertaining, enriching, and inspiring. *Cycles of Life: EXPLORING BIOLOGY* brings you the latest biological discoveries and theories. It offers a breathtaking view of the origin and nature of life, from the simplest single-celled forms to complex plants and animals and human beings. As an introductory biology course, *Cycles of Life: EXPLORING BIOLOGY* goes beyond describing the subject matter of biology. It examines the scientific method and considers both its promises and limitations. It answers many of our questions and poses new ones time after time, so that we are continually probing into the innermost secrets of life.

Course Themes

The designer, academic advisors, and producers of *Cycle of Life: EXPLORING BIOLOGY* have developed this course around nine principal themes:

1. Life is **dynamic** and continually **evolving** through time.
2. The living world exhibits both diversity and unity: organisms vary immensely, yet they also share many fundamental characteristics.
3. Life is a marvelously **complex** system of prolonging order, all sustained by **energy**.
4. There is **continuity** in the living world: Life presents a continuous stream of genetic information passed on through the ages.
5. The natural world exhibits **organization**: the different structures we see in nature tend to repeat themselves in different organisms and on different scales.
6. Life seeks **stability**: organisms work to maintain a stable internal environment despite changing external conditions.
7. Life is both **interdependent** and **competitive**: all organisms depend on the environment and one another for their basic needs, yet they must struggle against nature and each other to survive.
8. In biology, as in all the sciences, we must compare explicit **hypotheses** and predictions to **evidence** obtained from observations and experimental data.
9. Because biology is a **human endeavor**, it is subject to both the limitations and the inspirations of those who study it.

Course Components

Video Programs

Like most college courses, *Cycles of Life: EXPLORING BIOLOGY*, includes a textbook, student study guide, assignments, and tests. Because *Cycles of Life: EXPLORING BIOLOGY* is a *tele*course, however, it includes an additional learning element that traditional college courses do not have: a set of companion half-hour video programs, one for each of the 26 lessons in the course.

The 26 video programs feature leading practitioners, theoreticians, and academicians, in fields ranging from anatomy to zoology, who describe and explain fundamental concepts of our living world. Using human experience as a focus for this exploration, the video programs will take you to places you have never been and show you sights you could never see without the aid of micro-videography or computer simulation. Using footage of actual laboratory experiments and interviews with biologists, microbiologists, and biochemists, the video programs will also give you insights into the application of biological principles in research and development.

Textbook

The textbook for this telecourse is *Biology: Concepts and Applications*, 4th edition, by Cecie Starr (Brooks/Cole Publishing Company). Instructors may also choose to assign *Biology: The Unity and Diversity of Life,* 8th edition, by Cecie Starr and Ralph Taggart (Brooks/Cole Publishing Company). The content of these textbooks parallels the content of the video programs.

Telecourse Student Guide

The telecourse student guide is your road map through *Cycles of Life: EXPLORING BIOLOGY*. It provides a starting point for each lesson by listing the reading, viewing, and doing activities for the lesson in a step-by-step fashion. It also includes an overview of the lesson's content and accompanying video program and a complete array of learning activities to help you master the learning objectives for the lesson.

Each lesson in this guide has the following components:

> **Assignments:** Detailed instructions on activities and reading assignments to be completed before and after viewing each video program.
>
> **Overview:** An introduction to the main topics covered in the textbook and video program.
>
> **Learning Objectives:** Statements of what you should learn from reading the textbook assignments, completing the activities in this guide, and viewing the video programs.

Viewing Notes: A capsule description of the video program for each lesson plus questions for you to consider while watching the program. The Viewing Notes are *not* a substitute for actually watching the video program.

Review Activities: Matching and Completion exercises to help you review and reinforce your understanding of important terms and concepts.

Self-Test: A brief multiple-choice quiz that allows you to test your understanding of the material in the lesson.

Using What You've Learned: Suggestions for additional activities to enhance your understanding and knowledge of biology. (Your instructor will indicate how you are to incorporate these exercises into your studies and which ones to do, and he or she may designate them as assignments for extra credit.)

The end of this student guide contains an answer key section. This section provides solutions to the Matching and Completion exercises and answers for the Self-Test questions. Check your answers after you have completed each activity. If you have any incorrect answers, review the material.

How to Take a Telecourse

If you are new to college courses in general, and to telecourses in particular, you can perhaps profit from a few suggestions offered by students who have successfully completed other telecourses.

Telecourses are designed for busy people—people with full-time jobs or family obligations—who want to take a course at home, fitting the study into their own personal schedules. To complete a telecourse successfully, you will need to plan in advance how to schedule your viewing, reading, and study. Buy the textbook, telecourse student guide, and any other materials required by your instructor before the course begins and look them over; familiarize yourself with any materials supplied by your college and estimate how much time it will take you to complete special tests and assignments for each lesson. Write the dates of midterms, finals, review sessions, and special projects on your calendar so that you can plan to have extra time to prepare for them. You may find it enjoyable and instructive to watch the programs with other people, but save the talking and discussion until after the programs so that you won't miss important information. After the program, take a few minutes to write a brief summary of what you have seen and the meaning of key concepts and terms, then answer the questions listed at the end of the Viewing Notes for each lesson.

The following suggestions about how to study and complete *Cycles of Life: EXPLORING BIOLOGY* successfully have been compiled from students who have satisfactorily completed telecourses.

- *Do* buy both the text and the telecourse student guide for *Cycles of Life: EXPLORING BIOLOGY* or arrange to share copies with a friend. Do *not* try to get through this course without these books.

- *Do* watch each of the video programs. In order to pass the examinations, you will need to read and study the text and to view the programs. If you have a videocassette recorder, tape the programs for later review.

- *Do* keep up with your work for this course every week. Even if you do not have any class sessions on campus or any assignments to turn in, you should read the text and do the assignments in the student guide, as well as watch the video programs. Set aside viewing, reading, and study time each week and stick to your schedule.

- *Do* get in touch with the faculty member who is in charge of *Cycles of Life: EXPLORING BIOLOGY* at your college or university. The instructor can answer any questions you have about the material covered in the course. Your faculty member can also help you catch up if you are behind, advise you about additional assignments, discuss the type of test questions you can expect, and tell you where you might be able to watch programs you have missed or wish to review.

- *Do* complete all the Review Activities and Self-Tests provided in this guide. These will help you master the Learning Objectives and prepare for formal examinations.

- If you miss a program or fall behind in your study schedule, don't give up. Many television stations repeat broadcasts of the programs later in the week or on weekends. Your college might have videocassette copies of programs available in the campus library or media center. And *do* call on your course faculty member or manager to help if you have problems of any kind. This person is assigned specifically to help you succeed in this telecourse.

Acknowledgments

Producing the *Cycles of Life: EXPLORING BIOLOGY* telecourse was a complex team effort by many skilled people. Several of the individuals responsible for this course are listed on the copyright page of this book. In addition, special thanks are extended to Brent G. DeMars, Ph.D., assistant professor of biology at Lakeland Community College (Ohio), who reviewed the content and accuracy of this second edition of the telecourse student guide.

Cycles of Life: EXPLORING BIOLOGY was produced in cooperation with Brooks/Cole Publishing Company, NILRC (Northern Illinois Learning Resources Consortium), Oregon Community College Distance Education Consortium, New Jersey Community College Telecommunications Consortium, and Texas Consortium for Educational Telecommunications.

The production of *Cycles of Life: EXPLORING BIOLOGY* was guided by a national academic advisory committee for the telecourse. Committee members proposed the telecourse themes and objectives, reviewed the video programs as they were developed, and provided innumerable valuable suggestions on content and resources for the video programs.

NATIONAL ADVISORY COMMITTEE MEMBERS

Jerry Button, Portland Community College (Oregon)

Jack Carey, Biology Editor, Brooks/Cole Publishing Company

Jim Danielson, University of Nebraska, Lincoln

Nancy Dengler, University of Toronto

Brian Earle, Cedar Valley Community College (Texas)

Samuel Huang, Riverside Community College (California)

Gary Karpen, The Salk Institute, La Jolla, California

Carolyn Robertson, Tarrant County Junior College (Texas)

Cecie Starr, Textbook Author, Brooks/Cole Publishing Company

Tim Tirrell, Cumberland County College (New Jersey)

The 26 programs of *Cycles of Life: EXPLORING BIOLOGY* were produced in the studios of KOCE-TV. An affiliate of the Public Broadcasting Service (PBS), KOCE-TV is owned and operated by the Coast Community College District. This study guide and other materials for this course were developed for publication by the Office of Instructional Systems Development at Coastline Community College (Fountain Valley, California), a member of the Coast Community College District.

Contents

Preface v

Lesson 1: Biological Concepts 1
Lesson 2: Chemical Foundations 13
Lesson 3: Cell Structure and Function 25
Lesson 4: Metabolism 39
Lesson 5: Energy In—Energy Out 52
Lesson 6: Mitosis and Meiosis 64
Lesson 7: Patterns of Inheritance 76
Lesson 8: DNA Structure and Function 90
Lesson 9: Proteins 103
Lesson 10: Microevolution 116
Lesson 11: Macroevolution 129
Lesson 12: Viruses, Bacteria, and Protistans 142
Lesson 13: Fungi, Plants, and Animals 154
Lesson 14: Plants: Tissues, Nutrition, and Transport 168
Lesson 15: Plants: Reproduction and Development 181
Lesson 16: Animals: Structure and Movement 193
Lesson 17: Animals: Circulation 208
Lesson 18: Animals: Immunity 222
Lesson 19: Animals: Respiration 235
Lesson 20: Animals: Digestion and Fluid Balance 248
Lesson 21: Animals: The Neural Connection 263
Lesson 22: Animals: Endocrine Control 276
Lesson 23: Animals: Reproduction and Development 290
Lesson 24: Populations and Communities 303
Lesson 25: Ecosystems and the Biosphere 314
Lesson 26: The Human Factor 328

Answer Key 343

LESSON 1

Biological Concepts

Assignments

For the most effective study of this lesson, we suggest that you complete the assignments in the sequence listed below:

Before Viewing the Video Program

- Read the Overview and Learning Objectives for this lesson. Use the Learning Objectives to guide your reading, viewing, and thinking.

- Read Chapter 1, "Concepts and Methods in Biology," pages 2–18, in the Starr textbook.

    Or read, in the Starr/Taggart textbook, Chapter 1, "Concept and Methods in Biology," pages 2–18.

- Read the Viewing Notes in this lesson.

View "The Unity and Diversity of Life" Video Program
After Viewing the Video Program

- Briefly note your answers to the questions at the end of the Viewing Notes.

- Review all reading assignments for this lesson, especially the Chapter 1 Summary on page 16 in either textbook and the Viewing Notes in this lesson.

- Write brief answers to all the Review Questions at the end of Chapter 1 in the textbook to be certain you understand the text material.

- Complete the Review Activities in this guide to reinforce your understanding of important terms and concepts. Check your answers with the Answer Key and review when necessary.

- Take the Self-Test in this guide to measure your achievement of the Learning Objectives. Check your answers with the Answer Key and review when necessary.

- Complete the Using What You've Learned activities and any other activities and projects assigned by your instructor.

Overview

From the fundamental characteristics that define life to the diversity of higher organisms and their complex interactions, Lesson 1 surveys the field of biology to lay the conceptual foundations for your study of the *Cycles of Life*. As you proceed through the lesson, you will confront some of the same questions humankind has asked for centuries: *What is life? What drives the instinct to survive? What are the common characteristics that link all living things? What accounts for the diversity of life that surrounds us? How do parents pass on their genetic blueprint to their offspring?*

In Lesson 1, you will learn about some of the collective wisdom to date with respect to these questions and others. You will learn about the central role of energy in supporting the life of individual organisms and the complex ecosystems in which they live. You'll learn how reproductive processes provide a way of maintaining continuity from one generation to the next, while allowing for genetic alterations that help species adapt to their changing environment. You'll discover how the incredible diversity of life on this planet masks a common ancestry and biochemical similarities that persist to this day. Your search for answers will only begin in this lesson; it will continue throughout subsequent lessons as you develop a greater understanding of the *Cycles of Life*.

Because this telecourse places special emphasis on the everyday work of biologists and other life science professionals, Lesson 1 introduces the methodological foundations that underlie scientific investigation. Through the conceptual explanations and real-life examples presented in your course materials, you'll begin to understand what motivates researchers such as Dr. Paul Saltman and Dr. Christopher Wills. You'll learn how hypotheses are generated and tested in a painstaking effort to add to our understanding of the natural world. You'll learn about theories, explanations of phenomena that are based on systematic observation, not internal conviction.

One theory that has stood the test of time is Darwin's theory of evolution. It is based on the concept of natural selection, the process by which characteristics appear and disappear in populations over time, based on whether they help or hinder individual members in their struggle to survive. Not only does the theory of evolution explain the diversity of life that has characterized the planet, it stands as a testament to the scientific process and the usefulness of theories in helping us understand the world around us.

Learning Objectives

When you have completed all assignments for this lesson, you should be able to:

1. List the features that distinguish living organisms from nonliving matter.
2. Describe the general pattern of energy flow through the Earth's life forms, and explain how their interactions help cycle the Earth's resources.
3. Summarize how DNA can affect the traits of offspring from generation to generation.
4. List and generally describe the five *kingdoms* into which living organisms are classified.
5. Explain what is meant by the term *diversity*, and identify possible causes for the great diversity of life forms on Earth.
6. Explain what is meant by the term *unity*, and identify possible causes of similarities among Earth's organisms.
7. Discuss briefly how scientists came to believe that the populations of organisms that inhabit Earth have evolved through time.
8. Generally describe how biologists proceed through a scientific investigation.
9. Distinguish between a scientific *hypothesis* and a *theory*.
10. Identify some limitations imposed on science and scientists.

Viewing Notes

To help you understand these "Biological Concepts," the video program for Lesson 1 provides a broad visual overview of life on this planet and how science has helped us to understand and appreciate its complexity. You'll meet Dr. Paul Saltman as he attempts to define the concept of life in the biological sense and explain those characteristics that are shared by living organisms everywhere. The discussion helps you understand the central role of energy in the life cycle of organisms and explains how they obtain and process energy through the chemical processes of metabolism.

From Dr. Christopher Wills, you'll learn about another process which is essential to the survival of living organisms: reproduction. He explains the importance of reproduction to the survival of individual cells and their multicellular "hosts."

In Part Two of the program, you'll investigate the basic process scientists use to learn more about the world around us. Your hosts for this segment are Professor Bonnie Roohk and Dr. John Moore. Through their explanations, you'll learn about the basic steps of the scientific method and the importance of each step in building a universal base of biological knowledge. Moreover, you'll learn about the fundamental attitudes toward curiosity, objectivity, and truth that drive the scientific method and the efforts of scientists such as Professor Roohk and Dr. Moore. You'll also learn about the nature of theories and how they help us understand the natural world.

In the third and final segment of the program, you'll examine one of the most familiar—and important— "theories" in biology, the concept of evolution. With Dr. Blaire Van Valkenburgh, you'll travel back in time to examine the fossil record and trace the evolution of prehistoric wolves and saber-toothed cats to their modern descendants. In the process, you'll learn how scientists use fossils and other clues to reconstruct the natural history of times long past and identify the forces that led to the diversity of life that surrounds us today. Throughout Dr. Van Valkenburgh's presentation, you'll gain a greater appreciation of the ingenuity and dedication that have driven developments in the field of biology and continue to characterize its progress.

As you watch the video program, consider the following questions:

1. What are some of the characteristics shared by all living things?
2. Why is energy so important to living organisms?
3. What is metabolism and what is its purpose?
4. How is life perpetuated from one generation to another?
5. How does sexual reproduction help a species adapt to a changing environment?
6. What are the principal steps in the scientific method?
7. What is a hypothesis, and what is its importance to scientific investigation?

8. How would a scientific investigator define the concept of "truth"?
9. How does the fossil record help us understand our natural world today?
10. How do Darwin's concepts of evolution and natural selection help explain the diversity of life on Earth today?

Review Activities

Matching

Match the terms listed below with the definitions that follow. Check your answers with the Answer Key and review any terms you missed.

I.

____ 1. experiment   ____ 4. scientific method
____ 2. hypothesis   ____ 5. test
____ 3. prediction   ____ 6. theory

a. a proposed explanation for a particular phenomenon

b. a testable explanation of a series of related phenomena

c. a test in which a certain phenomenon is manipulated in controlled ways in an effort to better understand its nature

d. an effort to create actual observations that match predicted or expected observations

e. an expectation of a particular observation based on the assumption that a given theory or hypothesis is correct

f. a systematic approach for acquiring knowledge about phenomena; involves the formulation of a problem, the collection of observable data, and testing of hypotheses

II.

____ 1. diversity   ____ 5. metabolism
____ 2. evolution   ____ 6. natural selection
____ 3. homeostasis   ____ 7. photosynthesis
____ 4. inheritance   ____ 8. reproduction

a. the acquisition and conversion of sunlight energy into chemical energy

b. the genetic transmission of structural and functional patterns from parent to offspring

c. the process by which traits and the individual organisms which inherit them become more or less prevalent in the population due to the adaptive advantages the traits provide

d. the process by which new generations of organisms are produced

e. the state in which the physical and chemical conditions of the organism's internal environment are being maintained within normal limits

f. the changes in physical and behavioral traits that occur within a lineage or lineages over time

g. the variations in form, function, and behavior that have accumulated among living organisms

h. all of the enzyme-mediated chemical reactions through which cells obtain and use energy

III.

___ 1. cell
___ 2. community
___ 3. consumers
___ 4. decomposers
___ 5. ecosystem
___ 6. organism
___ 7. population
___ 8. producers
___ 9. species

a. a group of individuals of the same species which occupy a given area

b. a collection of organisms and their physical environment that interact through the exchange of energy and materials

c. one or more populations of sexually reproducing organisms that can interbreed under natural conditions

d. the populations of all species occupying a particular habitat

e. the smallest unit of life which can survive and reproduce on its own

f. organisms that obtain energy from the physical environment

g. organisms that obtain energy and raw materials through the tissues of other organisms

h. organisms that obtain energy by the chemical breakdown of the remains, products, or wastes of other organisms

i. any entity that is capable of carrying on the activities of life

Completion

Fill in each blank with the most appropriate term from the list for that paragraph. A term may be used once, more than once, or not at all. If a question requires two or more answers in succession, they may be in any order (unless the question indicates otherwise). Check your answers with the Answer Key and review when necessary.

1. Life on Earth is characterized by both _____ and _____. All living organisms are similar in that they are made up of the same _____ and function according to the same _____ laws. They survive by processing energy through _____, by responding to conditions in the _____, and by reproducing through the genetic blueprint contained in their _____. Living organisms also exhibit great _____ in the way they obtain energy, in the way they interact with their _____, and the specific _____ they have evolved to adapt to conditions in the _____.

   matter   unity
   metabolism   reproduction
   natural selection   traits
   physical   ATP
   environment   DNA
   diversity   genes
   substances   offspring

2. Darwin's theory of _____ by _____ explains how the five _____ of life and the various _____ within each developed. The theory holds that individuals of a given _____ show different versions of the same _____, and that some of these _____ may affect the ability to _____ to _____. The most adaptive _____ become more common, while the least adaptive _____ become less common and eventually disappear from the _____.

   reproduce   evolution
   reproduction   gene(s)
   artificial selection trait(s)
   natural selection   grow
   population   survive
   species   biosphere
   kingdom(s)   ecosystem

Lesson 1 / Biological Concepts

3. _____ usually begins with the identification of a _____ or a _____. A _____ is proposed to try and answer the _____. From the _____, the investigator tries to make predictions about what can be expected in nature, and then tests these _____ through _____. These _____ enable the investigator to compare actual _____ with predicted _____. From the results of _____, the investigator draws _____ about whether the _____ should be accepted or rejected.

hypothesis	question
theories	principles
predictions	conclusions
convictions	observations
concepts	scientific inquiry
problem	variables
experiments	

Self-Test

Select the one best answer for each question. Check your answers with the Answer Key and review when necessary.

1. Living matter differs from nonliving matter in that living matter
   a. is responsive to external stimuli.
   b. goes through a cycle of growth and decay.
   c. contains a genetic blueprint for reproduction.
   d. consumes energy to drive its activities.

2. Which of the following describes the general pattern of energy flow among living organisms?
   a. consumers → producers → decomposers
   b. decomposers → consumers → producers
   c. producers → decomposers → consumers
   d. producers → consumers → decomposers

3. DNA is critical to reproduction because it
   a. ensures that offspring will have similar traits to the parents.
   b. produces offspring with traits that are radically different from the parents.
   c. allows offspring to inherit variations in traits from the parents.
   d. a and c

4. Which of the following kingdoms includes only single-celled organisms?
   a. Monera
   b. Protista
   c. Fungi
   d. Plantae

5. Which of the following best explains why life on Earth is so diverse?
   a. Variations in inherited traits persist or disappear based on their adaptive advantages.
   b. The genetic blueprint is rearranged from one generation to the next.
   c. Interbreeding between different species produces new traits.
   d. Natural disasters lead to the spontaneous origin of new lineages.

6. What is one of the reasons for the similarity among various life forms?
    a. They use the same form of energy to drive their internal processes.
    b. They are composed of the same basic chemical substances.
    c. There are only a limited number of possible genetic combinations.
    d. All of the above are reasons.

7. Which of the following provides scientists with the most compelling evidence to support the theory of evolution?
    a. the way in which living organisms are distributed over the Earth's surface
    b. comparing the actual appearance and disappearance of traits over time with mathematical projections
    c. study of fossils of long-extinct plants and animals
    d. observations of single-celled organisms over successive generations

8. The scientific process usually begins with
    a. a hypothesis proposed to explain a particular phenomenon.
    b. a prediction about how a phenomenon works.
    c. a test to determine the nature of a phenomenon.
    d. observations of a phenomenon under controlled conditions.

9. A theory differs from a hypothesis in that a theory
    a. can change as new evidence is accumulated.
    b. cannot be tested experimentally.
    c. is more certain than a hypothesis.
    d. encompasses a range of related hypotheses.

10. One of the limits of scientific inquiry is that
    a. some hypotheses cannot be objectively tested.
    b. objective observations are often open to interpretation.
    c. sometimes experimental results fail to support the hypothesis.
    d. the theories upon which hypotheses are based can change.

Using What You've Learned

Based on your own interests or your instructor's requirements, complete one or more of the following activities.

1. Is the term *scientific creationism* an oxymoron? Investigate this belief system and determine whether or not it qualifies as a science. Does a belief in evolution necessarily exclude a belief in religion? Why or why not?

2. Visit a particular habitat (a pond, a grassy area, a garden) and try to identify the different kinds of organisms you might find. Then, using a chart or diagram, show how these organisms contribute to the energy flow and cycling of materials in this area.

3. Find one or two articles on a new diet, exercise method, or medication. Look for references to studies done to validate the claims made, and determine if enough evidence has been provided for you to accept or reject the recommendations.

4. Go to the library and select a research article describing an experimental study. Identify the independent, dependent, and control variables. What are the purposes of these variables?

5. Investigate some of the ways that scientists study the biological world today. Select one technique and report on how it works. How does the technique contribute to the body of scientific knowledge?

Challenge Questions

1. How do biologists distinguish between the living and the nonliving?

2. What is metabolism? How does your metabolism compare to that of a plant? What are the similarities? The differences?

3. What are some examples of homeostasis? Why is it important for organisms to achieve homeostasis?

4. Why is it important for a species to be able to change, to "evolve"? What would happen if a species produced identical offspring in each successive generation?

5. What are the essential biological roles fulfilled by producers, consumers, and decomposers?

6. How was Darwin's development of the principle of evolution an example of the scientific method in action?

7. What are some examples of evolution occurring in nature or in the laboratory today?

8. How can you define truth in the context of scientific investigation?

9. How is a principle different from a belief? How is a theory different from a principle?

LESSON 2

Chemical Foundations

Assignments

For the most effective study of this lesson, we suggest that you complete the assignments in the sequence listed below:

Before Viewing the Video Program

- Read the Overview and Learning Objectives for this lesson. Use the Learning Objectives to guide your reading, viewing, and thinking.

- Read Chapter 2, "Chemical Foundations for Cells," pages 20–33, and Chapter 3, "Carbon Compounds in Cells," pages 34–49, in the Starr textbook.

  Or read, in the Starr/Taggart textbook, Chapter 2, "Chemical Foundations for Cells," pages 20–35, and Chapter 3, "Carbon Compounds in Cells," pages 36–53.

- Read the Viewing Notes in this lesson.

View the "Chemical Foundations of Life" Video Program

After Viewing the Video Program

- Briefly note your answers to the questions at the end of the Viewing Notes.

- Review all reading assignments for this lesson, especially the Chapter 2 and Chapter 3 Summaries on pages 32 and 48 in the Starr textbook (on pages 33 and 52 respectively in the Starr/Taggart textbook) and the Viewing Notes in this lesson.

- Write brief answers to all the Review Questions at the end of Chapters 2 and 3 in the textbook to be certain you understand the text material.

- Complete the Review Activities in this guide to reinforce your understanding of important terms and concepts. Check your answers with the Answer Key and review when necessary.

- Take the Self-Test in this guide to measure your achievement of the Learning Objectives. Check your answers with the Answer Key and review when necessary.

- Complete the Using What You've Learned activities and any other activities and projects assigned by your instructor.

Overview

Your study of the *Cycles of Life* will depend on a basic understanding of the building blocks of matter, organic matter in particular. These building blocks include subatomic particles—the electrons, protons, and neutrons that make up the atom, which in turn is the basic unit of matter from which molecules are made. How these particles interact with each other is the basis of chemistry, the principal theme of Lesson 2.

Together, the textbook readings and video program for Lesson 2 will help you understand the fundamentals of chemistry, including how and why atoms combine to form the increasingly complex molecules that are fundamental to life. You'll learn about the properties of water, the universal solvent so abundant on our planet, and its role in the survival of living organisms. You'll learn about acids, bases, and salts and how they support metabolism. Finally, you'll learn about the four "molecules of life"—the carbohydrates, proteins, lipids, and nucleic acids that play critical roles in so many biological processes.

Your exploration will bring you in contact with some of the life-sustaining substances we take for granted: simple and complex sugars, starches, and fatty acids; hormones, amino acids, and enzymes; hemoglobin and the nucleotides that are the chemical basis for DNA.

In the process of acquiring this foundation of chemical concepts, you'll learn about some of the tools and methods that scientists use to study chemical compounds and their behavior. You'll also learn about some of the fascinating ways that chemicals are used as tools to study other aspects of our world, such as the use of isotopes to date fossils, monitor the effects of air pollution, and treat disease.

Your study of chemistry will only begin, not end, in Lesson 2. In subsequent lessons, you will build on the basic foundation presented here to guide your further exploration of the *Cycles of Life*.

Learning Objectives

When you have completed all assignments for this lesson, you should be able to:

1. Define the term *matter*, and distinguish among an atom, a molecule, an element, and a compound.
2. Describe how protons, electrons, and neutrons are arranged into atoms.
3. Define the terms *atomic symbol, atomic number,* and *mass number (or "atomic weight")*, and use these concepts to describe the structure of selected elements that are significant to living things.
4. Define *isotope*, and identify possible uses for isotopes in scientific study.
5. Explain how electrons are distributed in atoms, and how this affects the number and types of chemical bonds that can be formed.
6. Describe the various types of chemical bonds, the circumstances under which each of them forms, and their relative strengths.
7. Identify the properties of water that make it vital for life.
8. Define the terms *organic* compound and *inorganic* compound, and give examples of each.
9. Understand how small organic molecules can be assembled into large macromolecules by *condensation* and how large macromolecules can be broken apart into their basic subunits by *hydrolysis*.
10. List the four large "molecules of life," identify their composition and structure, and describe some of the functions they perform in the cell.

Viewing Notes

The video program for Lesson 2 will introduce you to the "Chemical Foundations of Life," the molecules, atoms, and subatomic particles that make up living and nonliving matter. Computer-animated sequences will allow you to see what no human being has ever witnessed firsthand—the composition and interaction of atoms that is the basis for all chemical reactions. You'll learn about the delicate balance between protons, neutrons, and electrons, and why atoms have a tendency to gain or lose these subatomic particles to create electrically charged ions or atomic isotopes.

Isotopes are of particular interest to biologist Mark Poth, who is studying the effects of air pollutants on mountain forests. Because pollutants carry more isotopes than found in nature, measuring the isotope content of chemicals normally absorbed by trees provides an indication of how much they're affected by pollution.

The behavior of isotopes also illustrates how atoms combine to form compounds. You'll learn about three kinds of chemical "bonds": ionic bonds, covalent bonds, and polar bonds, and how these bonds enable living things to assemble complex organic compounds from the simple elements found in nature.

One of the major chemical constituents of life, water, is profiled in Part Two. Water is a simple compound consisting of two atoms of hydrogen bonded to one atom of oxygen. While much of our world is covered with water, most of it is too salty for human use.

As you explore the chemical properties of water, you'll meet Chief Engineer Gary Snyder, who will explain how desalination plants are used to turn sea water into fresh water. While the technology remains costly, you'll learn about some of the options being considered to make desalination a practical approach to meeting the planet's fresh water needs.

In Part Three, "Solving Molecular Mysteries," you will follow the progress of Dr. Susan Taylor in her quest to determine the structure of certain organic compounds and how they interact. In the process, you'll learn about the four groups of compounds crucial to life: carbohydrates, lipids, nucleic acids, and proteins.

As you focus on the nature of Dr. Taylor's detective work, you'll learn more about proteins and the building blocks of proteins, amino acids. By witnessing a computer-generated protein in the making, you'll discover how the sequence of constituent amino acids and the shape of the protein determine its function.

Dr. Taylor will also explain the role of one group of proteins—kinases—on regulating the activity of enzymes in the cell. Their significance will become apparent as you learn of the various diseases caused by defects in these protein kinases. You'll learn about the role X-ray crystallography played in establishing the structure of kinases and the mechanisms by which such diseases occur.

As you watch the video program, consider the following questions:

1. What is the basic structure of an atom? How are subatomic particles typically arranged?
2. Why do atoms interact with other atoms to form compounds? What attracts one atom to another?
3. How do ionic bonds, covalent bonds, and polar bonds differ?
4. What are isotopes, and what do they tell us about the chemical composition of living matter?
5. What are the chemical properties of water?
6. How are the properties of water exploited during the desalination process?
7. What is X-ray crystallography and how does it help us determine the structure of chemical compounds?
8. What are the basic chemical components of carbohydrates, proteins, lipids, and nucleic acids?
9. What kinds of tools and technologies are being used to identify the structure and function of these "molecules of life?"
10. How do amino acids determine the structure and function of a protein?
11. What is a kinase? What are the roles of kinases in the human body?

Review Activities

Matching

Match the terms listed below with the definitions that follow. Check your answers with the Answer Key and review any terms you missed.

I.

____ 1. atom
____ 2. chemical bond
____ 3. covalent bond
____ 4. electron
____ 5. element
____ 6. functional group
____ 7. hydrogen bond
____ 8. ion
____ 9. molecule
____ 10. neutron
____ 11. organic compound
____ 12. proton

a. the bond that results when an atom interacts weakly with a neighboring atom that is already part of a polar covalent bond

b. a unit of matter which is composed of atoms of the same or different elements which are chemically bonded

c. an atom or compound which loses or gains an electron, thereby acquiring a positive or negative electrical charge

d. the smallest unit of an element

e. a negatively charged unit of matter that orbits around the atomic nucleus

f. a unit of matter that has mass but does not have an electric charge

g. any carbon-based compound

h. an atom or group of atoms that bonds with an organic compound and influences its behavior

i. the union between two or more atoms or ions

j. a substance that cannot be broken down into component substances with different properties

k. the bond that results when electrons are shared between atoms or groups of atoms

l. a positively charged particle found in the atomic nucleus

II.

____ 1. alcohol ____ 6. nucleic acid
____ 2. amino acid ____ 7. polypeptide
____ 3. carbohydrate ____ 8. polysaccharide
____ 4. fatty acid ____ 9. protein
____ 5. lipid ____ 10. triglyceride

a. an ester of glycerol that contains three ester groups and involves one or more acids

b. general term for a simple sugar or large molecule consisting of sugar units

c. an organic compound consisting of one or more chains of amino acids held together by peptide bonds

d. an organic compound consisting primarily of carbon and hydrogen that dissolves readily in nonpolar solvents

e. an organic compound consisting of numerous covalently bonded sugar units

f. a small organic molecule consisting of a hydrogen atom, an amino group, an acid group, and an R group bonded to a central carbon atom

g. an organic compound with a long, flexible hydrocarbon chain and a –COOH group at one end

h. the long, single- or double-stranded chain of nucleotides that are joined by their phosphate groups

i. a hydrocarbon derivative with a hydroxyl (-OH) functional group

j. an organic compound consisting of three or more amino acids joined by peptide bonds

Completion

Fill in each blank with the most appropriate term from the list for that paragraph. A term may be used once, more than once, or not at all. If a question requires two or more answers in succession, they may be in any order (unless the question indicates otherwise). Check your answers with the Answer Key and review when necessary.

1. All matter is composed of one or more _____. Each _____ is composed of _____ with a unique number and arrangement of subatomic particles. These subatomic particles include _____, _____, and _____. Together, positively charged _____ and uncharged _____ make up the nucleus of the atom. Negatively charged _____ orbit the nucleus in energy shells or _____. When an atom gains or loses an electron, an electrically charged _____ is formed. Atoms with additional neutrons are called _____.

atoms	isotopes
compounds	molecules
electrons	neutrons
element(s)	orbitals
ion(s)	protons

2. When two or more _____ bond together, the result is a _____. _____ that consist of more than one kind of atom are called _____. Atoms enter into bonds with other atoms to fill vacancies in their electron orbitals. They do this by accepting a(n) _____ donated by another atom (_____) or by sharing _____ with another atom (_____). When each atom in a _____ exerts a different pull on the shared electrons, the result is a(n) _____. Water is an example of a compound held together by a _____. The polarity of water makes it an excellent _____, one which readily dissolves acids (substances that _____ hydrogen ions) and bases (substances that _____ hydrogen ions).

atoms	neutron
combine with	polar bond(s)
compounds	proton
covalent bond(s)	release
electron(s)	solvent
ionic bond(s)	solute
molecule(s)	

Lesson 2 / Chemical Foundations

3. The bonding versatility of _____ makes it the basis for _____ with their complex, three-dimensional shapes. The four principal types of _____ found in living organisms are _____, _____, _____, and _____. _____ are used as structural materials and as a source of energy. They are composed of simple or complex sugars. _____ include fatty acids and triglycerides, the body's richest source of energy. Amino acids are the main constituents of _____ which perform a variety of functions including structural support, molecular transport, and regulation. _____ are the basis for DNA and the genetic code.

carbon	nucleic acids
carbohydrate(s)	organic compounds
hydrogen	polypeptide chains
hydrocarbons	polysaccharides
lipids	proteins
molecules	water

Self-Test

Select the one best answer for each question. Check your answers with the Answer Key and review when necessary.

1. The smallest unit of matter that cannot be broken into constituent substances with different properties is called a(n)
   a. compound.
   b. molecule.
   c. element.
   d. isotope.

2. Which of the following would you expect to find in the nucleus of an atom?
   a. electrons and protons
   b. electrons and neutrons
   c. protons and neutrons
   d. electrons, protons, and neutrons

3. Carbon, a key element in organic compounds, has an atomic number of six. This means that
   a. it can bond with six different atoms at the same time.
   b. it has six protons in the nucleus.
   c. it has six orbitals.
   d. it has an atomic weight equivalent to six units.

4. Isotopes are useful in scientific research because they
   a. indicate the level of radioactivity in an organism.
   b. can be used to speed up chemical reactions.
   c. will not interact with atoms that we wish to study.
   d. tell us how much of a given substance is from natural sources.

5. Atom A will tend to interact with Atom B if
   a. the outer orbitals of both atoms are filled.
   b. the outer orbitals of both atoms are unfilled.
   c. both atoms are electrically neutral.
   d. both atoms are electrically charged.

6. A polar covalent bond is one in which
   a. electrons are transferred from one atom to another.
   b. electrons between participating atoms are exchanged.
   c. electrons are shared equally by participating atoms.
   d. electrons are shared, but drawn more to one atom than another.

7. The behavior of the hydrogen bonds in water molecules enables living organisms to
   a. exclude the use of water in biochemical reactions.
   b. regulate internal temperature.
   c. keep their cells from bursting.
   d. use water as a buffer for acid-base interactions.

8. An organic compound is one that
   a. only occurs in nature.
   b. contains carbon atoms.
   c. is only used in biochemical reactions.
   d. cannot be broken into separate molecules.

9. Large molecules are often broken to provide energy through the process of
   a. condensation.
   b. electron transfer.
   c. functional-group transfer.
   d. hydrolysis.

10. Which of the following "molecules of life" are built from chains of amino acids?
    a. carbohydrates
    b. lipids
    c. proteins
    d. nucleic acids

Lesson 2 / Chemical Foundations

Using What You've Learned

Based on your own interests or your instructor's requirements, complete one or more of the following activities.

1. Using colored styrofoam balls and sticks, construct examples of simple atoms. Using the same materials, show how atoms use different types of chemical bonds to form simple molecules such as water, sodium chloride, ammonia, and so forth.

2. Combine water and salad oil in a jar. Shake the contents and observe the results. Is the resulting solution a compound or a mixture? Explain your conclusion.

   Now combine iced tea mix with sugar and water and observe the results. Is this solution a mixture or a compound? Again, explain your answer.

3. Using a coiled telephone cord with different colored bands, show the three structural levels of proteins. Stretch the cord out to show primary structure. Use the normal, coiled cord to show the helical secondary structure. Now fold the coiled cord into a three-dimensional shape to show the tertiary structure.

4. Bring in a variety of commercial food products and note the ingredients. What ingredients tend to show up most often? To what class of organic compounds do they belong? Why are these compounds so prevalent in commercial foods?

5. Go to the library and find a recent article on a drug, a poison, or a chemical pollutant that presents a health threat to human beings. Why is this pollutant a threat to health? By what chemical mechanism or mechanisms does it exert its effects? Prepare a report on your findings.

Challenge Questions

1. Why are isotopes of a given element found less frequently in nature than the basic structure of the element? How are isotopes formed?

2. Why do some atoms prefer to share electrons to form covalent bonds rather than donate them to form ionic bonds?

3. If like charges repel, how does the nucleus of an atom, with its positively charged protons, keep from flying apart?

4. What accounts for water's unique chemical and physical properties?

5. What is pH, and what does it measure?

6. Why do proteins form secondary, tertiary, and quaternary structures? What are the advantages of these structures in terms of biochemical reactions?

LESSON 3

Cell Structure and Function

Assignments

For the most effective study of this lesson, we suggest that you complete the following assignments in the sequence listed below:

Before Viewing the Video Program

- Read the Overview and Learning Objectives for this lesson. Use the Learning Objectives to guide your reading, viewing, and thinking.

- Read Chapter 4, "Cell Structure and Function," pages 50–73, in the Starr textbook. Also read Chapter 5, "Ground Rules of Metabolism," pages 84–88.

  Or read, in the Starr/Taggart textbook, Chapter 4, "Cell Structure and Function," pages 55–79, and Chapter 5, "A Closer Look at Cell Membranes," pages 82–91.

- Read the Viewing Notes in this lesson.

View the "Secrets of the Cell" Video Program

After Viewing the Video Program

- Briefly note your answers to the questions at the end of the Viewing Notes.

- Review all reading assignments for this lesson, especially the Chapter 4 Summary on pages 71–72 in the Starr textbook (Chapter 4 and Chapter 5 Summaries on pages 77–78 and 94 respectively in the Starr/Taggart textbook) and the Viewing Notes in this lesson.

- Write brief answers to all the Review Questions at the end of Chapter 4 in the Starr texbook (Chapter 5 in the Starr/Taggart textbook) to be certain you understand the text material.

- Complete the Review Activities in this guide to reinforce your understanding of important terms and concepts. Check your answers with the Answer Key and review when necessary.

- Take the Self-Test in this guide to measure your achievement of the Learning Objectives. Check your answers with the Answer Key and review when necessary.

- Complete the Using What You've Learned activities and any other activities and projects assigned by your instructor.

Overview

In the previous lesson, you learned about the chemical foundations that form the basis for organic material and life itself. In Lesson 3, you will learn how this material is organized into the essential building blocks of all living organisms: the cells.

Cells are highly sophisticated machines with lives of their own. While each cell has the basic tools and skills to sustain itself, the cells in your body and those of other complex organisms have evolved specialized functions to aid in tasks such as nerve transmission, movement, reproduction, and digestion. Thus we have neurons, muscle cells, egg and sperm cells, and the cells of the digestive tract.

Despite the diversity of form and function, we find a remarkable consistency in the internal structure and activities of these cells. All cells contain the genetic instructions to build the molecules of life. All cells have structures to transfer energy from the biochemical breakdown of organic material and structures to manufacture the substances that the cell itself is made of. All cells have mechanisms for moving molecules in and out of the cell and for enabling the cell to interact with its environment.

Your exposure to the chemical concepts presented in Lesson 2 will help you understand what is perhaps the most basic cell structure: the "plasma membrane." Through control of its own internal chemical composition, the cell makes it possible for some nutrients and waste materials to pass through this membrane. Materials that cannot be imported or exported passively across the membrane are actively "escorted" by transport proteins to their destination.

With this knowledge of the basic anatomy and physiology of cells, you will be ready to explore variations on the cellular theme. For example, there are important differences between the types of cells found in plants and animals (eukaryotic cells), and the cells that bacteria are made of (prokaryotic cells). The smaller size and simplified construction of prokaryotic cells makes them particularly adaptable and widespread in our world.

You will also learn about important differences between plant and animal cells. Plant cells, for example, have a cell wall surrounding the plasma membrane and a "central vacuole" that stores nutrients and toxic wastes. Plant cells need sunlight and carbon dioxide to provide metabolically useful energy, and one of the by-products of this process, oxygen, is a critical ingredient in the animal cell's cycle of life. Your exploration of these mechanisms will continue in the next two lessons, where you will examine more closely the metabolic processes of photosynthesis and cellular respiration.

Learning Objectives

When you have completed all assignments for this lesson, you should be able to:

1. Understand the basic tenets of *cell theory*, and identify the scientific contributions that led to its development.

2. Use the term *fluid mosaic model* to describe the general structure and function of a cell's membranes, and identify the importance of membranes as a structural component of cells.

3. Define the terms *solute*, *solvent*, and *solution*, and relate them to the concentration of a solution.

4. Describe *diffusion*, including *osmosis*. Explain what causes diffusion, and identify the factors that influence its rate.

5. Define and give examples of *hypertonic*, *hypotonic*, and *isotonic* solutions, and note the direction of solute and solvent movement when any two of these solutions are separated by a selectively permeable membrane.

6. Distinguish those mechanisms by which substances move across membranes passively (without the use of energy) from active mechanisms (that require energy).

7. Contrast the general features of *prokaryotic* and *eukaryotic* cells.

8. Describe the membranous and nonmembranous *organelles* found in eukaryotic cells, and explain the general function of each.

9. Identify those features that generally distinguish plant cells from animal cells.

10. Identify some of the variations that create diversity of cells between organisms or specialization within one organism.

Viewing Notes

The video program for Lesson 3 makes an important connection between the behavior of complex organisms and the behavior of the individual cells that compose these organisms. As you will discover, almost everything an organism does can be explained by the actions of these tiny fragments of life.

Part One uses a familiar analogy to illustrate how an organism functions under hostile conditions. Like the athlete in a football game, the individual cell has developed a set of physical characteristics and skills to survive and succeed at the game of life. These characteristics and skills, while basic to all cells, have been modified and refined to enable the cell to perform a specialized function.

Sophisticated computer animation is used to take you on a guided tour of a typical animal cell to view structures and processes that can only be seen with the aid of powerful electron microscopes. It is here that you will learn about the internal structures that constitute the living machine we call the cell.

In Part Two, "Permeable Packaging," you'll learn more about the composition of the cellular membrane and the importance of this structure to the cell and life in general. Plant physiologist Robert Health uses hydroponic experiments to explain the flow of nutrients through this selectively permeable membrane. The structure of the lipid bilayer and the movement of molecules across it through passive and active transport mechanisms are graphically illustrated, underscoring the concept of the cell membrane as a sophisticated gatekeeper.

In the final segment of the video program, "The Simplest Cells," microbiologist Dennis Focht contrasts the eukaryotic cells that constitute plants and animals with prokaryotic cells, such as bacteria. Drawing on historical events, ecological models, and medical science, Dr. Focht describes the tremendous impact these single-celled organisms have had and continue to have on human survival. You'll also discover the physiological basis for their incredible adaptability to the harshest conditions on this planet.

As you watch the video program, consider the following questions:

1. How do the physiological functions of human beings and other complex organisms reflect the basic functions of individual cells?

2. What are the fundamental internal structures of cells, and how do these structures result in the basic functions of the cell?

3. How are cells able to use similar internal structures to perform different, specialized tasks?

4. How is the cell able to import and export materials through the selectively permeable plasma membrane?

5. What are the mechanisms for packaging and actively moving molecules that would not otherwise penetrate this membrane?

6. How are plant cells and animal cells similar? How are they different?

7. What is the role of prokaryotic cells such as bacteria in the cycle of life?

8. How are the prokaryotic cells that characterize bacteria different from the eukaryotic cells that constitute plants and animals?

9. In what way do prokaryotic cells have an advantage over eukaryotic cells in the battle for survival?

Review Activities

Matching

Match the cell structures pictured below with the terms that follow. Some terms may be used more than once. Check your answers with the Answer Key and review any terms you missed.

I. Plant Cell

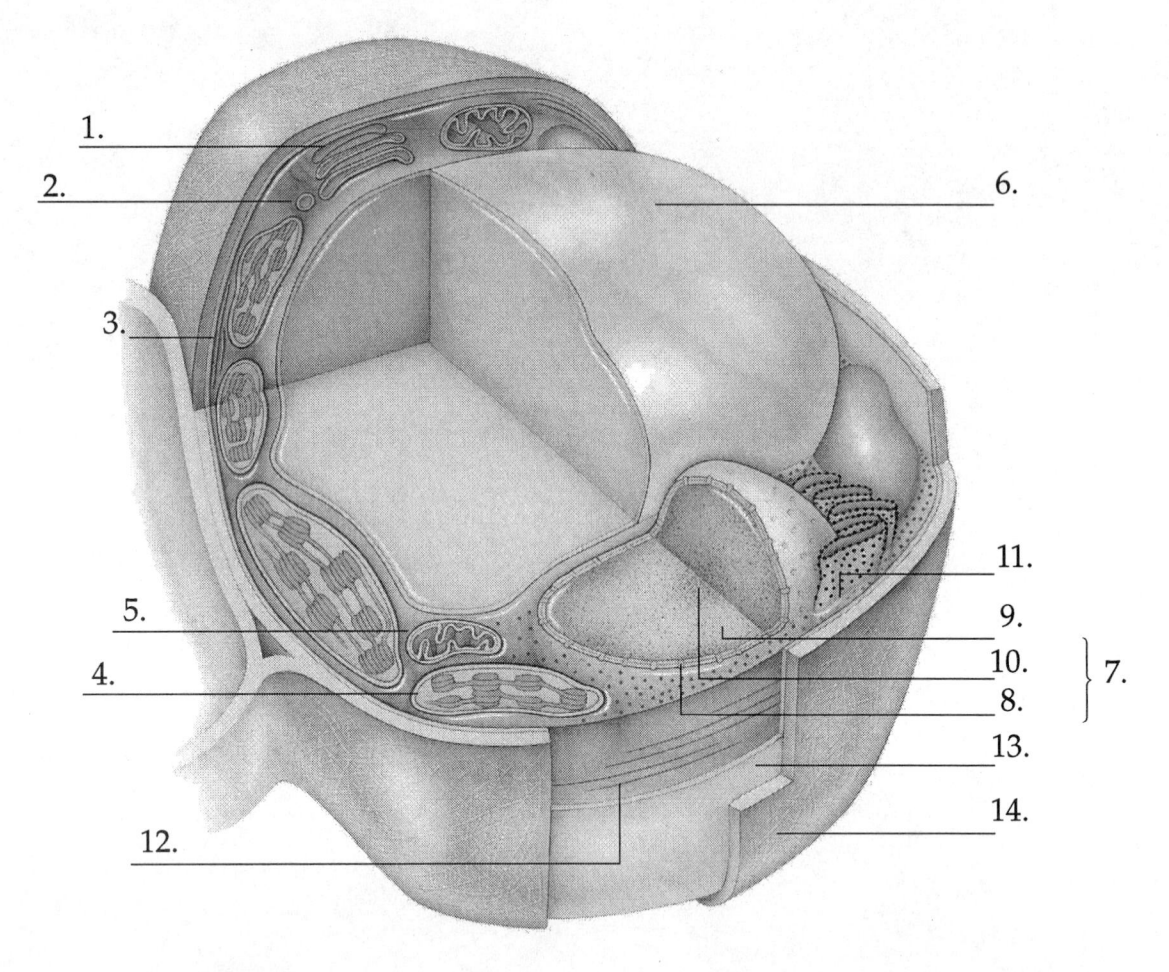

a. cell wall
b. central vacuole
c. chloroplast
d. DNA and nucleoplasm
e. microfilaments
f. endoplasmic reticulum
g. Golgi body
h. mitochondrion
i. microtubules
j. nucleolus
k. nuclear envelope
l. nucleus
m. plasma membrane
n. vesicle

30 Lesson 3 / Cell Structure and Function

II. Animal Cell

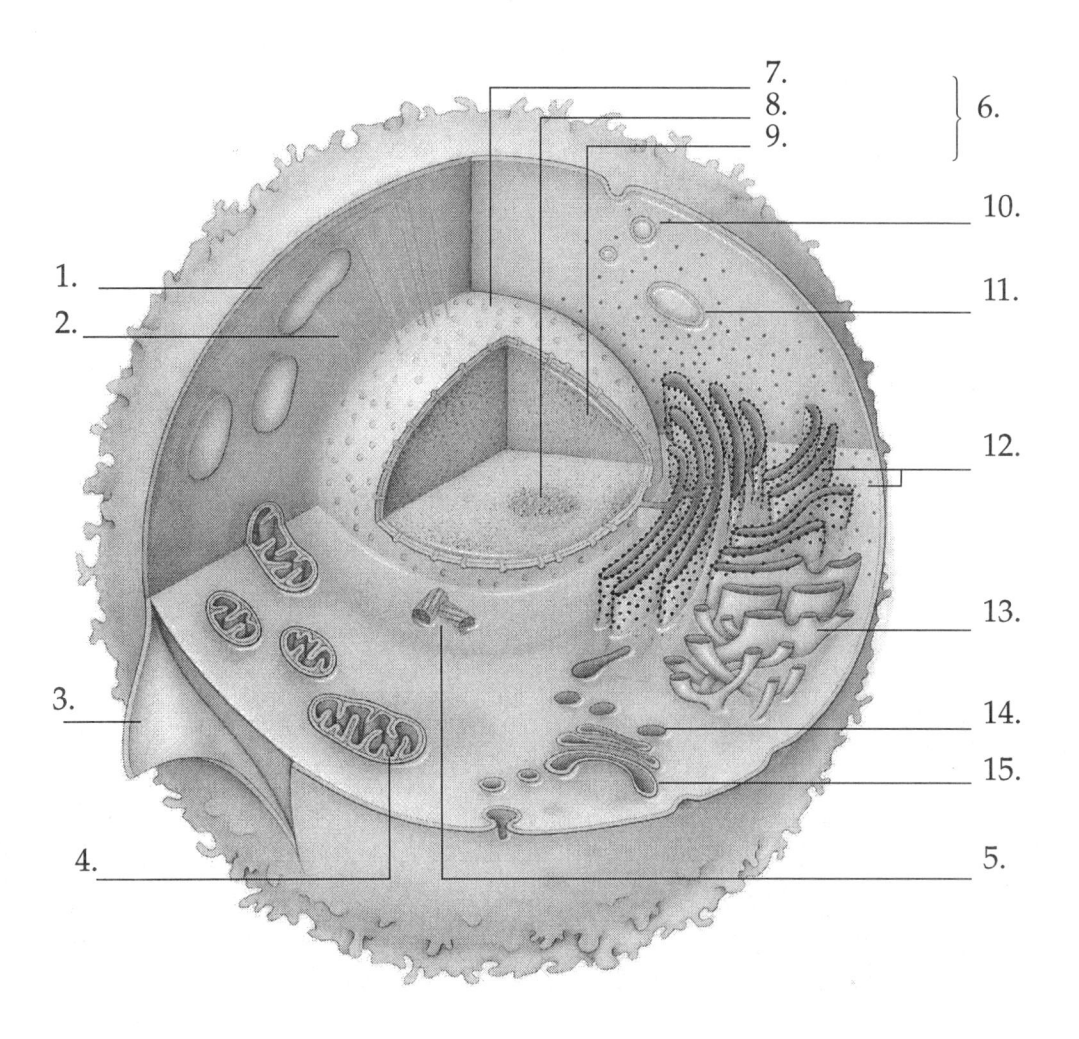

a. centriole(s)
b. endoplasmic reticulum, rough
c. endoplasmic reticulum, smooth
d. Golgi body
e. lysosome
f. microfilaments
g. microtubules
h. mitochondrion
i. nuclear envelope
j. nucleolus
k. nucleoplasm + DNA
l. nucleus
m. plasma membrane
n. vesicle

Lesson 3 / Cell Structure and Function

Now match the terms listed below with the definitions that follow. Check your answers with the Answer Key and review any terms you missed.

III.

___ 1. diffusion

___ 2. isotonic

___ 3. solvent

___ 4. active transport

___ 5. hypotonic

___ 6. osmosis

___ 7. solute

___ 8. passive transport

a. the protein-assisted pumping of a substance through a membrane against its concentration gradient

b. any substance dissolved in a solution

c. a condition in which two substances in a solution, separated by a membrane, have the same concentration

d. the net movement of like molecules in the direction of their natural concentration gradient, i.e., from higher to lower concentration

e. the unassisted movement of a substance through a membrane following the natural concentration gradient

f. a fluid in which one or more substances is dissolved

g. a solution of relatively lower solute concentration than another solution

h. the tendency of water to move from one side of a membrane to the other in response to its concentration gradient

Completion

Fill in each blank with the most appropriate term from the list for that paragraph. A term may be used once, more than once, or not at all. If a question requires two or more answers in succession, they may be in any order (unless the question indicates otherwise). Check your answers with the Answer Key and review when necessary.

1. The development of the _____ by Hooke followed by steady improvements in _____ enabled scientists to observe _____ firsthand. Their observations of plant and animal tissues led to the three basic tenets of the _____: a) that _____ organisms are composed of one or more cells and cell products, b) that the cell is the _____ unit having the properties of _____, and c) that life is perpetuated by the _____ and _____ of _____ cells.

all	life
animalcules	many
cells	micrograph
cell theory	organelles
compound microscope	organic matter
division	origin of life
electron microscope	single
growth	smallest
largest	spontaneous generation
lenses	staining techniques

2. The _____ provides a framework for understanding the structure of the cell _____. The membrane is composed of a _____, which contains primarily _____ and _____. Since solute molecules naturally move from a _____ solution to a _____ solution through the process of _____, the movement of nutrients greatly depends on its concentrations on both sides of the membrane. Specialized proteins called _____ help move nutrients against this natural tendency when needed by the cell.

carbohydrates	membrane
cell theory	messenger proteins
diffusion	mucosal barrier
endocytosis	osmosis
exocytosis	phospholipids
fluid mosaic model	proteins
hypertonic	solute
hypotonic	solvent
inorganic matter	transport proteins
lipid bilayer	wall

Lesson 3 / Cell Structure and Function

3. Cells are surrounded by a _____ membrane that protects their interior from the outside environment. In _____, this interior is composed of a material called _____, within which several membrane-bound compartments called _____ can be found. One of these organelles, the _____, contains the _____ for cellular reproduction. Other organelles perform specific functions. For example, the _____ produce energy for the cell, while _____ and the _____ assist in the synthesis of polypeptide chains. The Golgi body modifies these polypeptide chains into mature _____ for cellular use or secretion. Unlike eukaryotic cells, _____ such as bacteria are much simpler in construction. They do not have _____ to separate metabolic functions or a _____ to facilitate movement. Instead of using _____ and _____ to shift internal structures and produce movement, bacteria use _____ extending from their surface to move quickly through fluids.

animalcules	lysosome
centriole	microfilaments
cilia	microtubules
coelum	mitochondria
cytoplasm	mucosal
cytosine	nonpermeable
cytoskeleton	nucleolus
DNA	nucleus
endoplasmic reticulum	organelles
eukaryotic cells	plasma
flagella	prokaryotic cells
Golgi body	proteins
lipids	ribosome

Lesson 3 / Cell Structure and Function

4. While they are structurally similar, plant cells and animal cells are different in important respects. In addition to a plasma membrane, _____ cells have _____. Only plant cells have _____, which convert the energy of _____ to the energy of _____. Plant cells synthesize organic compounds from _____ and _____. Plant cells also have a _____, which stores amino acids, sugars, ions, and toxic wastes.

amyloplasts	glucose
animal	heat
carbon dioxide	mitochondria
cell wall	motion
central vacuole	nuclear envelope
centriole	oxygen
chemical bonds	plant
chloroplasts	proteins
chromatin	sunlight
cytomembrane system	water

Self-Test

Select the one best answer for each question. Check your answers with the Answer Key and review when necessary.

1. One of the basic tenets of cell theory is that
   a. all cells have the same internal structures.
   b. cells are the smallest units having the properties of life.
   c. cells are spontaneously generated from biochemical reactions.
   d. all cells use sunlight to produce energy in the form of ATP.

2. The cellular membrane is best described as a "mosaic" of proteins, lipids, and other substances that
   a. is one molecule in thickness.
   b. maintains equivalent fluid concentrations on both sides.
   c. is unusually rigid and impermeable.
   d. allows some but not all substances to pass in and out of the cell.

3. The concentration of a given solution depends on
   a. the atomic weight of the solute.
   b. the speed with which the solute is dissolved.
   c. the amount of solute dissolved in the solvent.
   d. the volume of the solution.

4. The rate at which a solute diffuses across a selectively permeable membrane depends on all of the following EXCEPT
   a. the electrical charge of the solute.
   b. the size of the cell.
   c. the concentration of the solute on either side of the membrane.
   d. the composition and structure of the membrane.

5. Solution A and Solution B contain the same solute and are separated by a membrane that allows the solute to pass freely. If Solution A is hypertonic relative to Solution B, you would expect
   a. the solute to move naturally from Solution A to Solution B.
   b. the solute to move naturally from Solution B to Solution A.
   c. no net movement of solute in either direction.
   d. proteins to move the solute against the concentration gradient.

6. The movement of nutrients through the lipid bilayer of the cell membrane against the concentration gradient is accomplished through

    a. passive transport.
    b. lysosomes.
    c. transport or carrier proteins.
    d. the microtubules of the cytoskeleton.

7. Prokaryotic cells differ from eukaryotic cells in that prokaryotic cells

    a. do not reproduce by cellular division.
    b. have a nucleus.
    c. use organelles to separate metabolic functions.
    d. are much smaller and more simply constructed.

8. The organelle that contains the DNA for cellular reproduction is the

    a. nucleus.
    b. mitochondria.
    c. Golgi complex.
    d. endoplasmic reticulum.

9. Unlike animal cells, plant cells use chloroplasts

    a. for cellular reproduction.
    b. to provide useful energy and organic compounds.
    c. to eliminate excess carbon dioxide.
    d. to store amino acids, lipids, and sugars.

10. Cells that have specialized functions, such as fat cells and muscle cells, are able to perform these functions because

    a. they have evolved new structures to accommodate these functions.
    b. they form cell "clusters" that behave differently from single cells.
    c. the nucleus "reprograms" the organelles of each cell as needed.
    d. they make greater or lesser use of the same basic internal structures.

Using What You've Learned

Based on your own interests or your instructor's requirements, complete one or more of the following activities.

1. Using a container of water and a small amount of colored ink or food dye, observe how molecules diffuse from areas of greater concentration to areas of lesser concentration. Use an eyedropper to place a few drops of the dye into the container and note how the dye diffuses through the water.

2. Design a hypothetical cell that would function best under each of the conditions listed below. Sketch or describe how the size, structure, and function of the cell membrane and organelles would be affected in each case.

    Condition A: Desert—high sunlight, high temperature, and extreme drought
    Condition B: Ocean bottom—saline, extreme pressure, and minimal light
    Condition C: High altitude—near weightlessness and minimal atmosphere

3. Visit the library and scan recent science publications for current research and scientific applications in the field of cellular physiology. Select the most interesting article and construct a brief report summarizing the principal outcomes.

4. Using the data presented in your textbook and common materials to construct scale models, compare the size of typical prokaryotic and eukaryotic cells. Then construct a scale model of a eukaryotic cell and its major organelles. How much of the cell's volume is occupied by these organelles?

5. Based on your knowledge of the human body and observations of other organisms in nature, try to list as many different cell types as you can. When you have completed the list, review each cell type and imagine how that cell's structure has been modified to accommodate this role.

Challenge Questions

1. How do the physiological functions of complex organisms reflect the basic functions of individual cells?

2. How are cells able to perform specialized tasks?

3. How does the structure of a cell reflect its function? How do the cell's internal organelles illustrate this connection between form and function?

4. What mechanisms can the cell use to import and export materials through the selectively permeable plasma membrane?

5. What are the important differences between eukaryotic cells and prokaryotic cells? Between plant cells and animal cells?

6. What is the role of prokaryotic cells such as bacteria in the cycle of life?

LESSON 4

Metabolism

Assignments

For the most effective study of this lesson, we suggest that you complete the following assignments in the sequence listed below:

Before Viewing the Video Program

- Read the Overview and Learning Objectives for this lesson. Use the Learning Objectives to guide your reading, viewing, and thinking.

- Read, in the Starr textbook, Chapter 5, "Ground Rules of Metabolism," pages 74–91. Although pages 84–88 were assigned in Lesson 3, a review of these pages will aid in your understanding of concepts presented in this lesson.

  Or read, in the Starr/Taggart textbook, Chapter 6, "Ground Rules of Metabolism," pages 96–111.

- Read the Viewing Notes in this lesson.

View "The Power of Metabolism" Video Program

After Viewing the Video Program

- Briefly note your answers to the questions at the end of the Viewing Notes.

- Review all reading assignments for this lesson, especially the Chapter 5 Summary on pages 89–90 of the Starr textbook (Chapter 6 Summary on page 110 in the Starr/Taggart textbook) and the Viewing Notes in this lesson.

- Write brief answers to all the Review Questions at the end of Chapter 5 in the Starr textbook (Chapter 6 in the Starr/Taggart textbook) to be certain you understand the text material.

- Complete the Review Activities in this guide to reinforce your understanding of important terms and concepts. Check your answers with the Answer Key and review when necessary.

- Take the Self-Test in this guide to measure your achievement of the Learning Objectives. Check your answers with the Answer Key and review when necessary.

- Complete the Using What You've Learned activities and any other activities and projects assigned by your instructor.

Overview

In Lesson 3, you learned what cells are made of and how the internal machinery of the cell actually works. In Lesson 4, you will continue this exploration, looking more closely at the chemical reactions that provide the cell with the energy it needs to carry out its various tasks. In the process, you'll begin to see that the intricate tapestry of life is actually woven from the thousands of chemical reactions that occur every moment, in every living organism. These are the processes of metabolism.

By the laws of nature, energy can neither be created nor destroyed, so living organisms have evolved elaborate mechanisms to draw the energy they need from different sources. For example, plants use chloroplasts to convert the sunlight into usable energy. Animals have developed other mechanisms for digesting food and extracting the energy stored in the chemical bonds of the nutrients. These metabolic processes all involve the exchange of molecules, atoms, and electrons according to the chemical principles explained in Lesson 2.

As you learn more about these exchanges, you will see patterns in the types of chemical reactions that make up metabolism. You will find that many of the reactions are "reversible," allowing the organism to keep the concentration of many substances in perfect balance. If a chemical reaction produces too much of a certain substance, the excess product triggers feedback mechanisms that reverse the reaction, bringing both source and product molecules back into balance. Chemical reactions that require small amounts of energy are often coupled with reactions that release small amounts of energy, providing an efficient mechanism for synthesizing needed substances from the breakdown of others. Some reactions are organized in elaborate linear or cyclical pathways designed to drain every ounce of usable energy from the progressive breakdown of complex organic compounds.

During your study of these processes, you will learn about a fascinating group of organic compounds called enzymes. These large, complex substances greatly increase the rate of metabolic processes. Enzymes bring the appropriate source materials into contact and encourage the formation of new chemical bonds. Because they remain unchanged through this process, enzymes can perform their matchmaking task over and over again. In Lesson 4, you'll learn how enzymes are able to accomplish this remarkable feat and what kinds of factors affect their performance.

With the knowledge gained from Lesson 4, you'll gain an appreciation for the role energy plays in sustaining life and for the mechanisms that have evolved to obtain this energy. You'll also have the background needed to understand more complex processes such as cellular respiration and cellular reproduction, the subjects of Lessons 5 and 6.

Learning Objectives

When you have completed all assignments for this lesson, you should be able to:

1. Summarize the two *laws of thermodynamics* that govern energy and its conversion from one form to another.

2. Identify the major reasons why cells use energy.

3. Describe the reorganization of atoms that occurs in a simple *reversible reaction* using equations and the principles of *equilibrium*.

4. Explain how *enzymes* speed up chemical reactions, and describe the mechanisms that regulate enzyme activity.

5. Describe the properties of *acids*, *bases*, and *buffers*. Offer an example of how each of them functions in the human body, and relate this to the concept of enzyme regulation.

6. Contrast the overall sequence of reactions in *linear* versus *cyclic* metabolic pathways, noting the roles of enzymes and cofactors (including some vitamins and minerals).

7. Describe the energy flow that occurs when energy-requiring reactions and energy-releasing reactions are coupled.

8. Use the ATP/ADP cycle as an example to explain how a cell can use a molecule to temporarily hold and transport energy for use in later reactions.

9. Describe some of the methods scientists use to investigate biochemical reactions.

Viewing Notes

"The Power of Metabolism," the video program for this lesson, draws on the experiences of Olympic gymnast Cathy Rigby-McCoy and Danish national women's tennis champion Merete Stockman, both of whom found the competitive edge to be a fine line separating health and illness. Their accounts illustrate the effects of starvation on human metabolism—the thousands of chemical reactions that occur in our bodies every moment.

The program shows how metabolism provides the energy the body needs to stay warm, to synthesize proteins, and to transmit nerve impulses. It also shows what happens when the body is deprived of its primary source of chemical energy—food—through eating disorders such as anorexia and bulimia.

With Dr. Paul Saltman as your guide, you'll actually witness the rearrangement of atoms that characterizes a metabolic reaction. In the process, you'll learn about the role of enzymes in speeding up these reactions. A computer simulation provides a closer look at the structure of a typical enzyme and how enzyme-catalyzed reactions really work.

In Part Two, "The Light from Within," Dr. Margo Haygood discusses how bioluminescence—organically produced light—provides a more vivid example of metabolic reactions. This phenomenon is important because it gives scientists a way of observing the underlying mechanisms of metabolic reactions and measuring their rate.

The third segment of the video program, "Fueling the Fire," reexamines the concept of energy in terms of the laws of thermodynamics and its importance in forestalling the natural tendency of living systems toward increasing entropy, or disorder. Professors Haygood and Saltman describe how "coupled reactions" control the release of energy and conserve it for constructive purposes. Such reactions are essential for building ATP, the principal, energy-rich molecule used by cells to drive metabolic processes.

The video program concludes with a closer look at metabolic reactions involving the cyclical synthesis and breakdown of ATP, reactions that ultimately drive the bodily functions that ensure our survival.

As you watch the video program, consider the following questions:

1. Why does the human body need energy? What forms of energy does the body require?
2. How does the body get the energy it needs to fuel its chemical reactions?
3. What does it mean when a reversible chemical reaction has reached "equilibrium"?
4. What is the role of enzymes in metabolism?
5. How do enzymes "speed up" the rate of chemical reactions?
6. What factors influence the behavior of enzymes?

7. How does the "bioluminescence" of certain types of life help scientists study metabolic reactions?
8. What are the first and second laws of thermodynamics?
9. Do these laws hold true for metabolic processes?
10. What is a "coupled" reaction?
11. What is ATP and why is it important to metabolism?

Review Activities

Matching

Match the terms listed below with the definitions that follow. Check your answers with the Answer Key and review any terms you missed.

I.

___ 1. activation energy   ___ 5. feedback inhibition

___ 2. coupled reaction   ___ 6. induced-fit

___ 3. entropy   ___ 7. metabolic pathway

___ 4. equilibrium   ___ 8. phosphorylation

a. the degree of disorder in a system

b. a control mechanism in which the production of a substance triggers a chain of events that inhibits its continued production

c. a sequence of enzyme-mediated reactions as found in living organisms

d. the pairing of an energy-releasing reaction with an energy-requiring reaction

e. model used to describe the way in which a substrate changes an enzyme's active site, thereby creating a better bond

f. the minimum energy required to start a reaction between colliding atoms

g. the attachment or transfer of an unbound phosphate group

h. the point at which a chemical reaction occurs in both directions (forward and backward) at the same rate

II.

___ 1. ATP
___ 2. acid
___ 3. base
___ 4. buffer
___ 5. catalyst
___ 6. coenzyme
___ 7. cofactor
___ 8. end product
___ 9. enzyme
___ 10. phosphate group
___ 11. substrate

    a. a molecule or group of molecules that an enzyme can chemically recognize, bind to, and modify in some way

    b. an organic compound that serves as the primary energy carrier in cells

    c. a type of protein that speeds up chemical reactions between specific substances

    d. any substance that releases hydrogen (H$^+$) ions in water

    e. the part of the ATP molecule that gives off energy when it breaks free

    f. an organic molecule or metallic ion that assists enzymes or transports atoms or electrons

    g. a substance that can stabilize pH by combining with and/or releasing hydrogen ions

    h. any substance that increases the likelihood or rate at which a chemical reaction occurs

    i. the substance present at the end of a metabolic reaction

    j. an organic compound that transfers hydrogen atoms and electrons from one site to another, for example, NAD$^+$

    k. any substance that releases hydroxide (OH$^-$) ions in water

Completion

Fill in each blank with the most appropriate term from the list for that paragraph. A term may be used once, more than once, or not at all. If a question requires two or more answers in succession, they may be in any order (unless the question indicates otherwise). Check your answers with the Answer Key and review when necessary.

1. The _____ states that _____ can neither be created nor destroyed, only changed in form. Since these conversions are never perfect, some of the energy is dissipated into unusable forms. According to the _____, this _____ the level of _____ in the universe. To maintain a high degree of organization, living organisms use the energy of _____ and/or _____ to build, store, and break down substances that they need for survival. This process is called _____.

   catalysis
   chemical bonds
   decreases
   energy
   entropy
   first law of thermodynamics
   heat

   increases
   life
   metabolism
   molecular movement
   second law of thermodynamics
   sunlight
   third law of thermodynamics

2. The types of chemical reactions that characterize metabolism may proceed in one direction, from _____ to _____, or in both directions. Such reactions are said to be _____. An example of a _____ reaction is when an _____, such as HCl, combines with a _____, such as NaOH:

$$HCl + NaOH \longrightarrow H^+ + Cl^- + Na^+ + OH^- \longrightarrow NaCl + H_2O$$

When _____ chemical reactions produce _____ and _____ at the same rate (the reaction reading from left to right occurs at the same rate as the reaction reading right to left), the reaction has reached _____. Reactions that *require* energy to occur are said to be _____, while those that *release* energy are considered _____. When the energy released by an _____ reaction provides the energy required for an _____ reaction, we have what is called a _____.

acid	equilibrium
base	exergonic
bilateral	metabolic pathway
buffer	oxidation-reduction
catalysis	products
coupled reaction	reactants
electron transport system	reversible
endergonic	spontaneous
entropy	

3. Coupled reactions are an important part of _____ such as the _____. _____ is the cell's medium of energy exchange, much like money. When a(n) _____ attached to a(n) _____ molecule breaks free or is transferred to another molecule, energy is _____. The energy that results from this _____ process can be stored or used to fuel other chemical reactions.

ADP	metabolic pathways
ATP	oxidation
ATP/ADP cycle	phosphate group
coenzyme	phosphorylation
consumed	protein
electrons	released
electron transport systems	respiration

Lesson 4 / Metabolism

4. Metabolic reactions are often catalyzed by _____. Enzymes _____ reaction rates by binding to a specific _____ and lowering the _____ necessary to start a reaction. According to the _____, the _____'s bonds are strained to fit the _____'s _____, which makes them easier to break and form into new bonds with other substances. Enzyme activity is affected by environmental conditions such as _____ and _____, as well as the presence of _____ and _____ that assist with the transfer of _____. Organized arrays of enzymes and coenzymes are an important part of the _____ found in chloroplasts and mitochondria.

activation energy	light intensity
active site	pH
coenzymes	phosphate groups
cofactors	polarity
electron transport system	protein
electron(s)	reaction threshold
enzyme(s)	slow down
induced-fit model	speed up
intermediate	substrate
ion	temperature
key-lock model	transition energy

Self-Test

Select the one best answer for each question. Check your answers with the Answer Key and review when necessary.

1. The first law of thermodynamics states that
       a. the level of entropy in the universe is decreasing.
       b. energy flows spontaneously from less complex forms to more complex forms.
       c. energy can neither be created nor destroyed.
       d. energy cannot be changed from one form to another.

2. Cells use energy to
       a. grow and reproduce.
       b. manufacture inorganic compounds.
       c. modify the external environment.
       d. do all of the above.

3. A reversible reaction is said to have reached equilibrium when
       a. the number of reactant molecules equals the number of product molecules.
       b. energy is no longer needed to continue the reaction.
       c. all product molecules have been converted to reactant molecules.
       d. the rate of the forward reaction equals the rate of the reverse reaction.

4. Which of the following explains why enzymes are so effective in speeding up metabolic reactions?
       a. They lower the activation energy needed for a reaction to occur.
       b. They are unaffected by changes in the cell's internal environment.
       c. A single enzyme can bind with many different substrates.
       d. They take on the properties of substrates and participate in their reactions.

5. If the pH of the stomach were made basic by ingestion of too many antacid tablets, pepsin, the dietary protein-digestion enzyme of the stomach, would
       a. work more efficiently.
       b. work less efficiently.
       c. work the same, regardless of pH.
       d. autodigest itself.

Lesson 4 / Metabolism   49

6. Cofactors such as NAD$^+$ and Fe^{++} help enzymes catalyze reactions by
    a. modifying the shape of the enzyme's active site.
    b. scouting for substrates that are compatible with the enzyme.
    c. turning the enzyme on or off as needed.
    d. transferring protons or electrons from one place to another.

7. In coupled reactions,
    a. the end product becomes the starting substance for the same reaction.
    b. the energy stored in the reactants is always equal to the energy stored in the products.
    c. coenzymes are always used to transfer electrons.
    d. the energy released from the breakdown of one type of molecule is used to build another type of molecule needed by the cell.

8. In the ATP/ADP cycle, energy is released to drive endergonic reactions through
    a. the conversion of ADP to ATP.
    b. the transfer of a phosphate group from ATP to another molecule.
    c. the decomposition of ATP into its original chemical components.
    d. the bonding of an ATP molecule with an ADP molecule.

9. Bioluminescence provides scientists a good opportunity to study metabolic reactions because
    a. they can measure changes in the light produced by such reactions.
    b. each enzyme produces a characteristic color when it reacts with a particular substrate.
    c. the substances that are needed to begin such reactions are easily identified.
    d. bioluminescent fish and other sea life are so plentiful.

Using What You've Learned

Based on your own interests or your instructor's requirements, complete one or more of the following activities.

1. Consider how energy is converted from one form to another in everyday activities such as cooking, doing laundry, and so forth. Are these conversions perfect? If not, where does the extra energy go?
2. Explain the metabolic dysfunction in those suffering from phenylketonuria (PKU). Review health-related literature to find other metabolic dysfunctions associated with inborn or acquired diseases.
3. Visit the library and scan recent science publications for current research and scientific applications in the area of metabolism. Select the most interesting article and construct a brief report summarizing the principal outcomes.
4. Demonstrate the induced-fit model of enzyme-substrate reactions by putting on a latex glove and observing how the insertion of your hand (substrate) changes the shape of the glove (active site).
5. Consider your body's response to excess acids and bases. What might cause such an imbalance? What kinds of medications are used to neutralize their effects? How are they able to do this? (Hint: Check the ingredients in over-the-counter antacids.)

Challenge Questions

1. Does the existence and continuation of "life" violate the second law of thermodynamics? Explain you answer.
2. What forms of energy does the human body require, and why?
3. How can the human body extract the energy it needs from a ham-and-cheese sandwich?
4. Why is the phosphorylation of molecules by ATP important to metabolism?
5. How does your intake of vitamins and minerals affect your metabolism?
6. How would enzyme activity be affected if you developed a high fever?
7. What are some of the effects of food deprivation on the human body?

LESSON 5

Energy In—Energy Out

Assignments

For the most effective study of this lesson, we suggest that you complete the assignments in the sequence listed below:

Before Viewing the Video Program

- Read the Overview and Learning Objectives for this lesson. Use the Learning Objectives to guide your reading, viewing, and thinking.

- Read Chapter 6, "How Cells Acquire Energy," pages 92–107, and Chapter 7, "How Cells Release Energy," pages 108–124, in the Starr textbook.

  Or read, in the Starr/Taggart textbook, Chapter 7, "Energy-Acquiring Pathways," pages 112–129, and Chapter 8, "Energy-Releasing Pathways," pages 130–146.

- Read the Viewing Notes in this lesson.

View the "Energy In—Energy Out" Video Program

After Viewing the Video Program

- Briefly note your answers to the questions at the end of the Viewing Notes.

- Review all reading assignments for this lesson, especially the Chapter 6 and Chapter 7 Summaries on pages 105–106 and 123 in the Starr textbook (Chapter 7 and Chapter 8 Summaries on pages 127–128 and 145 respectively in the Starr/Taggart textbook) and the Viewing Notes in this lesson.

- Write brief answers to all the Review Questions at the end of Chapters 6 and 7 in the Starr textbook (Chapters 7 and 8 in the Starr/Taggart textbook) to be certain you understand the text material.

- Complete the Review Activities in this guide to reinforce your understanding of important terms and concepts. Check your answers with the Answer Key and review when necessary.

- Take the Self-Test in this guide to measure your achievement of the Learning Objectives. Check your answers with the Answer Key and review when necessary.

- Complete the Using What You've Learned activities and any other activities and projects assigned by your instructor.

Overview

Lessons 2 through 4 explained the basic chemical processes that are the foundation of cellular metabolism. In Lesson 5, this knowledge will serve you well as you delve more deeply into the primary energy-acquiring and energy-releasing reactions of metabolism: photosynthesis and respiration.

Your first stop in the exploration of metabolic reactions is the ancient pathway of photosynthesis, the primary energy-acquiring pathway for self-nourishing autotrophs such as plants. As you learned in Lesson 3, the photosynthetic conversion of light energy to chemical energy occurs primarily in the chloroplasts of plant leaves. These light-dependent reactions produce ATP through two different pathways in an inner membrane of the chloroplast called the thylakoid membrane system. Later, in light-independent reactions, some of this energy is used to synthesize sugar phosphates, the building blocks of sucrose, cellulose, and starch.

With this basic biochemical foundation, you'll be able to understand some of the ways plants have modified the photosynthetic process to accommodate different climates and seasonal changes. You'll also begin to appreciate the tremendous impact plant life has on the world's climate as well as on the survival of heterotrophs such as ourselves.

As you move from the world of plants to the world of animals, you'll learn about another important metabolic pathway, the energy-releasing process known as aerobic respiration. Aerobic respiration begins in the cytoplasm with a process known as glycolysis, in which glucose is cleaved by enzymes into two pyruvate molecules, with a net gain of two ATP. Subsequent biochemical reactions in the cellular mitochondria produce two more ATP in the Krebs cycle. In the third and final stage of aerobic respiration, electron transport phosphorylation produces 32 more ATP, making it the most efficient energy-releasing pathway in the living world.

Your survey of these metabolic pathways concludes with a look at another common energy-releasing process called anaerobic respiration. Using glucose again as the principal substrate, the fermentation process upon which anaerobic respiration is based is the method commonly used by bacteria and protistans in oxygen-free settings. Although the pathway begins with glycolysis, yielding two ATP molecules, the glucose molecule is not completely broken down as in aerobic respiration. Instead, the products of glycolysis are recycled to form NAD^+, a coenzyme with a central role in breakdown reactions.

Not only do these reactions shed light on how living organisms use energy to sustain themselves, they provide insights into the relationships between our primitive ancestors and their modern-day descendants. The ubiquity of these reactions illustrate the connectedness of all living organisms, from the simple, single-celled bacteria to human beings.

Learning Objectives

When you have completed all assignments for this lesson, you should be able to:

1. Outline the general flow of energy and matter in the living world, noting the relationship between photosynthesis and aerobic cellular respiration that cycles carbon dioxide, oxygen, and water.

2. Describe the key steps of the light-dependent and light-independent reactions of *photosynthesis*, noting the raw materials required, the end products, and the location for each phase.

3. Describe basic leaf structure and the functional advantages it offers plants.

4. Compare *chemosynthesis* and photosynthesis as energy-acquiring processes.

5. Describe the three major stages of aerobic cellular respiration, noting the raw materials and products of each phase.

6. Distinguish between *aerobic* and *anaerobic* energy-releasing pathways, and identify the requirements and energy-producing efficiency of each process.

7. Understand how proteins and fats can be used as alternative energy sources in the aerobic respiration pathway.

8. Describe how the development of photosynthesis altered the evolution of life on Earth.

9. Describe some of the methods scientists use to study life's energy-acquiring and energy-releasing pathways.

Viewing Notes

The video program for Lesson 5 begins with an interesting visual metaphor: the Earth as a giant living cell, extracting energy from the sun to drive photosynthetic reactions, which in turn generate the chemical energy to support life in the form of plants, animals, and other organisms. This macroscopic perspective is balanced by the microscopic perspective presented in the subsequent segments of the program, where the biochemical pathways essential for life at the cellular level are examined in detail.

The first segment introduces Dr. Jeanne Erikson, who studies photosynthetic reactions of single-celled algae. She describes the passage of carbon dioxide and sunlight through the surface layer of plant leaves, where they become the source materials for the photosynthetic factories in the chloroplasts. Dr. Erikson explains how light triggers special light-trapping pigments to release energized electrons and how these electrons are passed through a series of biochemical reactions to produce NADPH and ATP. These light-dependent reactions are followed by the light-independent reactions of the Calvin-Benson cycle, where the six-carbon sugar glucose is ultimately formed.

Although oxygen is a mere by-product of these photosynthetic reactions, it is a crucial ingredient in aerobic respiration. Part Two describes this second process, which drives the metabolism of non-photosynthetic organisms such as ourselves. Dr. Paul Saltman explains how the chemical energy trapped in food particles is released through the chemical reactions of aerobic respiration, and how ATP plays a central role as the "currency" in the various energy deposits and withdrawals that characterize these transformations. Through computer animation, you'll witness the release of ATP through the series of chemical reactions that occur in the cellular cytoplasm and mitochondria: glycolysis, Krebs cycle, and electron transport phosphorylation.

Part Three of the video introduces yet another series of chemical reactions that have a significant role in metabolism—anaerobic respiration, or respiration *without* oxygen. From brewmaster Tom Burnes, you'll learn about the chemical basis of aerobic respiration, fermentation, and its role in beermaking. From Dr. Saltman, you'll learn about the role this process plays in the generation of energy for vigorous muscle activity.

As you watch the video program, consider the following questions:

1. Why is photosynthesis described in terms of light-dependent and light-independent reactions? How do these reactions differ?

2. How do the structure and composition of a plant leaf facilitate photosynthesis?

3. How do pigment molecules such as chlorophyll convert light energy into chemical energy?

4. What are the roles of carbon dioxide and oxygen in the photosynthetic process?

5. What are the main products of photosynthesis? Of glycolysis?

6. How much ATP is released through the process of aerobic respiration?

7. Where do the biochemical processes of aerobic respiration take place?
8. What is the purpose of anaerobic respiration, and what are the products?
9. How does anaerobic respiration help explain the cramps athletes feel after vigorous physical activity?

Review Activities

Matching

Match the terms listed below with the definitions that follow. Check your answers with the Answer Key and review any terms you missed.

I.

____ 1. aerobic respiration
____ 2. anaerobic electron transport
____ 3. autotroph
____ 4. C3 plant
____ 5. C4 plant
____ 6. Calvin-Benson cycle
____ 7. CAM plants
____ 8. fermentation
____ 9. glycolysis
____ 10. heterotroph
____ 11. Krebs cycle
____ 12. phosphorylation
____ 13. photosystem

a. an anaerobic process whereby glucose or another organic compound is broken down to produce pyruvate and two ATP

b. the principal energy-releasing metabolic pathway of ATP formation

c. the cyclic, light-independent reactions of photosynthesis

d. the stage in aerobic respiration in which pyruvate is broken down into carbon dioxide and, through coupling with electron transport phosphorylation, water

e. metabolic pathway in which a substance other than oxygen accepts electrons that have been removed from substrates

f. the attachment of unbound phosphate to a molecule

g. anaerobic pathway of ATP formation in which pyruvate from glycolysis is converted into lactate or ethanol

h. an organism that can build large organic molecules from carbon dioxide and energy from the surrounding environment

i. an organism that feeds on autotrophs, other heterotrophs or waste materials

j. clusters of light-trapping pigments in photosynthetic membranes

k. a plant that conserves water by opening stoma and fixing carbon dioxide at night

l. a plant that produces a three-carbon molecule, PGA, as the first intermediate of its carbon-fixing pathway; well-suited to hot climates

m. a plant that produces a four-carbon molecule, oxaloacetate, as the first intermediate of its carbon-fixing pathway, well-suited to cool climates

II.

___ 1. ATP
___ 2. ADP
___ 3. carotenoid
___ 4. chlorophyll
___ 5. FAD
___ 6. NAD
___ 7. PGA
___ 8. pigment
___ 9. pyruvate
___ 10. RUBP

a. a nucleotide coenzyme that delivers electrons and unbound protons from one metabolic reaction to another

b. a key intermediate compound in glycolysis and the Calvin-Benson cycle

c. a nucleotide that accepts an unbound phosphate to form ATP

d. a nucleotide that is the main energy carrier of the cell

e. a five-carbon compound that is used in carbon fixation during photosynthesis

f. a three-carbon compound that is one of the end products of glycolysis

g. a light-sensitive pigment molecule that absorbs light and donates electrons required for photosynthesis

h. a light-sensitive pigment molecule that transfers absorbed energy to chlorophyll

i. any light-absorbing molecule

j. a nucleotide coenzyme that is used to carry electrons and unbound protons during the fermentation process and the aerobic respiration process

Completion

Fill in each blank with the most appropriate term from the list for that paragraph. A term may be used once, more than once, or not at all. If a question requires two or more answers in succession, they may be in any order (unless the question indicates otherwise). Check your answers with the Answer Key and review when necessary.

1. As the primary pathway by which _____ and _____ enter the cycle of life, photosynthesis is crucial to the evolution and survival of life on Earth. _____ reactions inside the _____ of plants convert light _____ into the _____ energy of _____. With the energy of _____ and hydrogen and electrons from _____, _____ reactions in the chloroplasts' _____ produce the sugar phosphates used to synthesize _____, _____, and other end products of photosynthesis.

ATP	light-dependent
carbon	light-independent
carbon dioxide	NADPH
chemical bond	PGAL
chlorophyll	RUBP
chloroplasts	starch
energy	stroma
fats	sucrose
glucose	

2. Energy in the form of _____ bonds delivered by _____ provide the fuel for metabolic reactions. The primary energy-releasing pathway for multicelled organisms is _____, which consists of three stages. The first stage is _____, during which enzymes break down _____ to form _____ pyruvate molecules as well as _____ NADH and _____ ATP. In the second stage, the _____, cyclic reactions load coenzymes with hydrogen and electrons, forming _____ ATP and _____ carbon dioxide molecules for each pyruvate. In the third stage, _____ produces energy in the inner membrane of _____. Altogether, these three pathways generate _____ ATP for each _____ molecule metabolized.

28	glucose
36	glycolysis
48	Krebs cycle
aerobic respiration	lactate
ATP	mitochondria
Calvin-Benson cycle	NADH
chloroplasts	phosphate
electron transport phosphorylation	pyruvate
fermentation	three
five	two
four	

Self-Test

Select the one best answer for each question. Check your answers with the Answer Key and review when necessary.

1. The main pathway by which carbon and energy enter the living world is
   a. photosynthesis.
   b. respiration.
   c. decomposition.
   d. excretion.

2. In the light-dependent pathway of photosynthesis, the energy of sunlight is
   a. converted to ATP energy.
   b. used to manufacture glucose.
   c. used to split water into hydrogen and oxygen.
   d. used to catalyze the reaction between carbon dioxide and glucose.

3. Which of the following is a product of the light-independent pathway of photosynthesis?
   a. carbon dioxide
   b. glucose
   c. ATP
   d. all of the above

4. Why do leaves tend to have a large surface area?
   a. to provide an extensive network of thylakoid membranes
   b. to maximize the number of chloroplasts exposed to sunlight
   c. because chloroplasts are usually quite large
   d. they must provide a storage area for accumulated glucose

5. Chemosynthesis differs from photosynthesis in that chemosynthesis
   a. is an energy-releasing process.
   b. is much more efficient than photosynthesis.
   c. can occur in the absence of sunlight.
   d. doesn't occur in plants.

6. Which of the following is the first stage of aerobic cellular respiration?
   a. Krebs cycle
   b. electron transport phosphorylation
   c. glycolysis
   d. Calvin-Benson cycle

7. Anaerobic energy-releasing pathways differ from aerobic energy-releasing pathways in that anaerobic pathways
   a. use fermentation rather than glycolysis to split glucose.
   b. always generate alcohol as a metabolic by-product.
   c. are much more efficient for producing energy.
   d. do not use oxygen as the final acceptor of electrons.

8. Your body is able to use proteins and fats as alternative energy sources by
   a. using them as building blocks for glucose synthesis.
   b. activating alternative metabolic pathways that are not normally used.
   c. using special enzymes that change their molecular composition.
   d. breaking them down into intermediate products that can be used in aerobic respiration.

9. In what way did photosynthesis have a major impact on the evolution of life on Earth?
   a. The accumulation of oxygen as a photosynthetic by-product became the fuel for new aerobic life forms.
   b. The disappearance of carbon dioxide made nitrogen fixation possible.
   c. All life ultimately evolved from photosynthetic plants.
   d. Photosynthesis is the only way in which living organisms can obtain the energy for metabolism.

10. Scientists can study the energy-acquiring pathways and energy-releasing pathways best by
    a. watching metabolic reactions under an electron microscope.
    b. varying the light and chemical conditions in which organisms operate and observing the results.
    c. observing genetically engineered, single-celled organisms.
    d. developing theoretical models based on fundamental chemical principles.

Using What You've Learned

Based on your own interests or your instructor's requirements, complete one or more of the following activities.

1. Select one of the metabolic pathways discussed in this lesson and prepare a chart illustrating the sequence of biochemical events that constitute the pathway. Describe the required energy and/or source materials and the products in each step of the pathway.

2. Investigate the impact of atmospheric pollutants on the greenhouse effect and the implications for biochemical processes such as photosynthesis. Discuss the long-range impact of these changes and what is being done to curb the harmful effects of such pollutants.

3. Visit a greenhouse or nursery and inquire about the various strategies used to promote plant health and growth. Relate these strategies to the biochemical needs of plants.

4. Investigate one of the body-building programs available, including diet, exercise, and fluid intake regimens. Biochemically speaking, how does the program produce the effects it claims to produce? Do these claims seem justified?

5. Explore the process of alcoholic fermentation by visiting a winery or brewery and speaking with the winemaker or brewmaster. How does the process they use compare to the fermentation that occurs in your body? What kinds of substances do they use to trigger the fermentation process? How do they control the process?

Challenge Questions

1. Which light wavelengths promote photosynthesis? Which wavelengths inhibit it? What are the implications of this for growing plants under artificial conditions?

2. Which conditions would be best for plant growth—sunny and dry, or dark and moist? What are the tradeoffs involved in each case?

3. Why doesn't your body use nucleic acids as an energy source?

4. Why is it that some people are intoxicated or incapacitated after eating carbohydrate-rich food?

5. What is the "set-point theory" of metabolism, and what is the significance of this theory to people trying to lose or gain weight?

6. Why did living organisms evolve such elaborate mechanisms for obtaining ATP? Why not just obtain ATP directly from the environment?

7. How is it that some animals such as flour beetles and clothes moths can survive in areas where exogenous water is virtually unavailable?

8. By what mechanism does hydrogen cyanide—the lethal gas used in gas chambers—cause death?

LESSON 6

Mitosis and Meiosis

Assignments

For the most effective study of this lesson, we suggest that you complete the assignments in the sequence listed below:

Before Viewing the Video Program

- Read the Overview and Learning Objectives for this lesson. Use the Learning Objectives to guide your reading, viewing, and thinking.

- Read Chapter 8, "Cell Division and Mitosis," pages 126–137, and Chapter 9, "Meiosis," pages 138–151, in the Starr textbook.

  Or read, in the Starr/Taggart textbook, Chapter 9, "Cell Division and Mitosis," pages 148–159, and Chapter 10, "Meiosis," pages 160–173.

- Read the Viewing Notes in this lesson.

View the "Generations: Mitosis and Meiosis" Video Program
After Viewing the Video Program

- Briefly note your answers to the questions at the end of the Viewing Notes.

- Review all reading assignments for this lesson, especially the Chapter 8 and Chapter 9 Summaries on pages 136 and 150 in the Starr textbook (Chapter 9 and Chapter 10 Summaries on pages 158–159 and page 172 respectively in the Starr/Taggart textbook) and the Viewing Notes in this lesson.

- Write brief answers to all the Review Questions at the end of Chapters 8 and 9 in the Starr textbook (Chapters 9 and 10 in the Starr/Taggart textbook) to be certain you understand the text material.

- Complete the Review Activities in this guide to reinforce your understanding of important terms and concepts. Check your answers with the Answer Key and review when necessary.

- Take the Self-Test in this guide to measure your achievement of the Learning Objectives. Check your answers with the Answer Key and review when necessary.

- Complete the Using What You've Learned activities and any other activities and projects assigned by your instructor.

Overview

In Lesson 5, you investigated the complex chemical processes cells use to obtain and utilize energy to sustain life. Yet, for all their sophistication, these processes cannot ensure immortality. The survival of multicellular organisms, therefore, requires an ongoing mechanism for creating new cells.

In Lesson 6, you will find out how new cells are formed and how the instructions for complex processes such as respiration and photosynthesis are handed down from one generation to the next. You will learn about chromosomes, the cellular components that contain the cell's genetic record, and how their structure enables these records to be transmitted from parent to offspring through cellular division.

Much of Lesson 6 is devoted to the cellular reproductive processes of *mitosis* and *meiosis*. Mitosis is nature's version of cloning. It provides an asexual method for reproducing exact copies of a parent cell and is used in complex organisms for the replacement and growth of body cells. Meiosis, in contrast, is our insurance policy against environmental change. By shuffling the hereditary deck through the combination of genetic material from two parents, meiosis allows multicellular plants and animals to inherit new and varied traits. Some of these traits will provide a greater adaptive advantage than others.

Your survey of mitosis and meiosis will involve a microscopic examination of the intracellular changes that occur during cell division. Although these changes occur in predictable stages, there are important differences in what happens in the *somatic* cell during mitosis and the *germ* cell during meiosis. In meiosis, for example, germ cells undergo two divisions, typically producing four gametes, each with only half the chromosomes of the parent. In animals, the egg is the gamete that carries the maternal chromosomes, while the sperm is the gamete that carries the paternal chromosomes. It is only through fertilization —the joining of egg and sperm—that animal cells with the full complement of chromosomes can be born.

Your exploration of mitosis and meiosis will also involve a broader view of the implications of these processes on the survival of individual organisms and entire populations. Indeed, you will gain a better appreciation for the importance of these pivotal reproductive processes in perpetuating every species of plant and animal, including *Homo sapiens*.

Learning Objectives

When you have completed all assignments for this lesson, you should be able to:

1. Describe the basic features of chromosome structure and number in different organisms.
2. Understand what is meant by the *cell cycle*, and relate where *mitosis* fits into the cycle.
3. Summarize the events that occur during each phase of mitosis.
4. Explain how cytoplasm is divided between daughter cells after mitosis, contrasting the process in plant and animal cells.
5. Discuss the potential of the cell cycle to control events such as cancer formation and aging.
6. Contrast *asexual* and *sexual* types of reproduction that occur among unicellular and multicellular organisms.
7. Summarize the events that occur during each phase of *meiosis* and *gamete formation*.
8. Describe the events of meiosis and gamete formation that contribute to genetic variability among organisms.
9. Compare mitosis and meiosis with respect to overall purpose, where and when each occurs, and the resulting number of chromosomes.
10. Describe some of the methods and tools scientists use to study cell division.

Viewing Notes

In "Generations," the video program for Lesson 6, you will explore the process of cell division and its implications for the growth and survival of complex organisms such as ourselves. With the Great Exposition of 1851 as a backdrop, you will learn about important biological discoveries of the nineteenth century, including Walther Fleming's use of synthetic dyes to reveal the star players of cell division, the chromosomes.

With the help of Dr. Christopher Wills and Dr. Gary Karpen, you will learn about the significance of this discovery and how the structure of chromosomes is being gradually defined. It is this structure which provides the key to understanding the transmission of molecular instructions and inherited traits from one generation of cells to the next. From cytogenetics specialist Arlene Kumamoto, you'll learn how the chromosomes of different cells are routinely identified, matched, and counted using a laboratory procedure known as "karyotyping."

Part Two of the video program, "Mitosis and Merlot," will take you to a California vineyard to demonstrate how mitotic cell division is used to preserve the quality of specific grape varieties over successive generations. Through computer animation, you'll witness the process of mitosis, from the replication of DNA to its subsequent distribution.

Part Three contrasts the asexual process of mitosis with the sexual reproduction characteristic of meiosis. Citing an example of unsuccessful breeding, Dr. Kumamoto explains how zoologists were able to identify two different populations of the Dik-Dik antelope by the number of chromosomes present in the cells of each species. This case provides an appropriate springboard for discussing the specific mechanisms of meiosis as well as the advantages of sexual reproduction in general.

A combination of microscopy and computer animation reveals the genetic shuffling that occurs in meiosis and provides the basis for trait variations in offspring. Dr. Kurt Bernischke caps the program by explaining the significance of meiosis to the survival of life on Earth.

As you watch the video program, consider the following questions:

1. What are some of the methods that scientists have used to study the components, processes, and outcomes of cell division?

2. In what ways does the structure of the chromosome provide clues to the transmission of genetic information?

3. What is a centromere, and how does it contribute to chromosomal inheritance during cell division?

4. How does asexual reproduction by mitosis contribute to the growth and/or survival of the organism? What are the advantages and limitations of mitotic division as a method of reproduction?

5. How does sexual reproduction by meiosis contribute to the growth and/or survival of the organism? What are the advantages and limitations of meiotic division as a method of reproduction?

6. What are the steps in mitotic cell division, and how do they differ from the steps in meiotic cell division?

7. What kinds of mechanisms contribute to the mixing of genetic information during meiosis, and when do they occur?

8. What are gametes, and how do they produce a diploid cell with the normal chromosome number?

Review Activities

Matching

Match the terms listed below with the definitions that follow. Check your answers with the Answer Key and review any terms you missed.

I.

___ 1. asexual reproduction
___ 2. cell cycle
___ 3. chromosome
___ 4. fertilization
___ 5. gamete
___ 6. gene
___ 7. germ cell
___ 8. meiosis
___ 9. mitosis
___ 10. sexual reproduction

a. a type of cell whose primary function is the formation of gametes for sexual reproduction

b. a haploid cell that may combine with another haploid cell as during fertilization

c. a form of reproduction in which the offspring inherits the same genes as the parent

d. the interval between the time that new cells are produced and the time that those cells complete their own division

e. the type of cell/nuclear division responsible for gamete formation and sexual reproduction

f. a form of reproduction in which the genes from two different parents are combined in the offspring

g. a DNA molecule with attached proteins

h. the type of cell division which is used in asexual reproduction and tissue growth and repair

i. the union of two gametes

j. a specific portion of a chromosome that contains information for a particular inherited trait

II.

___ 1. allele
___ 2. cell plate
___ 3. centromere
___ 4. chromatid
___ 5. diploid cell
___ 6. egg
___ 7. haploid cell
___ 8. sperm
___ 9. spindle apparatus
___ 10. spore

a. half of a duplicated chromosome

b. haploid cell produced by plants for reproduction

c. any cell that has a pair of each type of chromosome

d. the region of the chromosome where you find attachment sites for microtubules which aid in nuclear division

e. an internal structure that pulls the chromosomes into the proper position for cell division

f. any molecular form of the same gene

g. the male gamete in animals

h. any cell with unpaired chromosomes

i. the female gamete in animals

j. in plants, a disklike structure where the wall that divides a parent cell into daughter cells is formed

Completion

Fill in each blank with the most appropriate term from the list for that paragraph. A term may be used once, more than once, or not at all. If a question requires two or more answers in succession, they may be in any order (unless the question indicates otherwise). Check your answers with the Answer Key and review when necessary.

1. Cells use the process of _____ to produce new generations of cells. While _____ cells such as bacteria reproduce by _____, _____ cells reproduce through _____ or _____. _____ divides the "parent" cell's chromosomes and the DNA they contain into two identical portions. The daughter cells have _____ chromosome number and characteristics that the "parent" cell had. _____ nuclear division _____ the number of chromosomes. The _____ formed in this process has _____ the parent cell's total number of chromosomes. It must fuse with another _____ to produce a cell with the appropriate chromosome number.

allele	halves
binary fission	haploid homologous chromosome
chromatid	meiosis
diploid	mitosis
doubles	micrographs
eukaryotic	oogenesis
fertilization	prokaryotic
gamete	reproduction
gene	the same
germ cell	twice
half	

Lesson 6 / Mitosis and Meiosis

2. _____ offers an advantage over _____ because it allows for unique combinations of genes in the offspring. The process of _____ consists of three events: _____, _____, and _____. _____ occurs in _____ cells called _____. During _____, chromosome segments _____ and homologous chromosomes are divided to produce _____ cells which may develop into _____. These mature _____ cells then fuse with other _____ from other parent cells, further shuffling the genetic material through the process of _____.

asexual reproduction	meiosis
cross over	mitosis
diploid	oogenesis
eggs	replicate
fertilization	sexual reproduction
four	sperm
gamete formation	spermatogenesis
gametes	spores
germ cells	two
haploid	

3. The _____ begins when a new daughter cell is born. During _____, the cell grows in size and complexity and duplicates its DNA in preparation for the reproductive phase of the _____. In _____, this reproductive phase begins when the _____ begin to _____ and the microtubules that form the cell's cytoskeleton begin to _____. This is _____. In the next phase, _____, the nuclear envelope _____ and _____ are aligned midway between poles of the completed _____. At _____, _____ separate and move to these poles. _____ marks their arrival at the spindle poles and the formation of new nuclear envelopes to surround the new sets of chromosomes.

anaphase	meiotic process
break(s) down	mitosis
cell cycle	mitotic process
centriole	metaphase
centromere	prophase
chromosomes	replicate
condense	telophase
interphase	sister chromatids
homologous chromosomes	spindle apparatus
meiosis	

Lesson 6 / Mitosis and Meiosis

Self-Test

Select the one best answer for each question. Check your answers with the Answer Key and review when necessary.

1. Which of the following statements is true for animal cells in general?
   a. They reproduce sexually.
   b. Daughter cells contain half as many chromosomes as the parent cell.
   c. They have a total of 46 chromosomes.
   d. During mitosis, newly duplicated chromosomes consist of two sister DNA molecules called chromatids.

2. Mitosis is the stage of the cell cycle during which
   a. haploid gametes are formed in preparation for sexual reproduction.
   b. the cell increases in size and complexity.
   c. nuclear division occurs.
   d. the cell's DNA is replicated.

3. Which of the following is the correct sequence of events in mitosis?
   a. anaphase —> interphase —> metaphase —> telophase
   b. prophase —> metaphase —> anaphase —> telophase
   c. anaphase —> metaphase —> interphase —> prophase
   d. telophase —> prophase —> metaphase —> anaphase

4. In animal cells, the cytoplasm divides
   a. when a cell plate forms at the cell's midsection.
   b. by pinching in two through a process called cleavage.
   c. after all the cell's organelles have been replicated.
   d. when repelling electrical charges push opposing spindles apart.

5. Cancer cells differ from normal cells in that
   a. cancer cells have an abnormal number of chromosomes.
   b. cancer cells inhibit the reproduction of other cells.
   c. cell division among cancer cells continues unchecked.
   d. cancer cells deposit toxins that poison the body.

6. A major difference between sexual and asexual reproduction is that in sexual reproduction, offspring cells
   a. can have different traits than those of the parent cell.
   b. are exact clones of the parent cell.
   c. have half as many chromosomes as the parent cell.
   d. have twice as many chromosomes as the parent cell.

7. The correct sequence of events in the sexual reproduction of animals is:
   a. fertilization —> gamete formation —> diploid cell —> meiosis
   b. meiosis —> diploid cell —> gamete formation —> fertilization
   c. gamete formation —> diploid cell —> fertilization —> meiosis
   d. diploid cell —> meiosis —> gamete formation —> fertilization

8. Chromosomal crossing over that contributes to genetic variability in the offspring first occurs in meiosis during
   a. prophase I.
   b. metaphase I.
   c. anaphase I.
   d. prophase II.

9. Daughter cells formed by meiosis have
   a. half the number of chromosomes as daughter cells formed in mitosis.
   b. twice as many chromosomes as daughter cells formed in mitosis.
   c. the same number of chromosomes as daughter cells formed in mitosis.
   d. exactly 46 chromosomes.

10. One of the methods scientists frequently use to study the process of cell division is
    a. capturing video footage through electron microscopes.
    b. growing cells taken from different organisms under laboratory conditions.
    c. observation of inherited traits resulting from controlled breeding.
    d. construction of computer models that simulate mitosis and meiosis.

Using What You've Learned

Based on your own interests or your instructor's requirements, complete one or more of the following activities.

1. Using posterboard, paper cut-outs, and some kind of tacking material, design a display that illustrates the processes of mitosis and/or meiosis. Show the various transformations that occur in cell division by manipulating the pieces in your display.

2. Design a similar display to contrast sexual reproduction in plants and animals.

3. Using different colored string or pop-it beads, demonstrate the concept of crossover.

4. Visit the library and scan recent science publications for current research on the role of the cell cycle in aging or in diseases such as AIDS or cancer. Select the most interesting article and construct a brief report summarizing the principal outcomes.

5. Design an experiment that would use the technique of breeding to track inherited traits over successive generations. How could this technique be used to isolate trait-specific alleles on the chromosome?

Challenge Questions

1. If cloning of human beings were possible, what would be the implications of asexual reproduction on human populations?

2. What are the three mechanisms that contribute to the enormous number of new gene combinations possible in sexual reproduction?

3. If a single bacteria reproduces itself every 15 minutes, how many bacteria cells will you find in a culture that began with one cell 8 hours ago?

4. How does cell division in plants differ from cell division in animals? Why have two different mechanisms evolved for dividing the cytoplasm?

5. How is it possible for *each* of two daughter cells to have the same number of chromosomes as the one parent cell?

6. How do anti-cancer drugs work? How do you think they affect the cell cycle?

7. How is it possible for some cells in the human body, such as skin cells, to continue to divide, while others, such as nerve cells, stop dividing after they reach maturity?

LESSON 7

Patterns of Inheritance

Assignments

For the most effective study of this lesson, we suggest that you complete the assignments in the sequence listed below:

Before Viewing the Video Program

- Read the Overview and Learning Objectives for this lesson. Use the Learning Objectives to guide your reading, viewing, and thinking.

- Read Chapter 10, "Observable Patterns of Inheritance," pages 152–169, and Chapter 11, "Chromosomes and Human Genetics," pages 170–189, in the Starr textbook.

    Or read, in the Starr/Taggart textbook, Chapter 11, pages 174–191, and Chapter 12, pages 192–215.

- Read the Viewing Notes in this lesson.

View the "Patterns of Inheritance" Video Program

After Viewing the Video Program

- Briefly note your answers to the questions at the end of the Viewing Notes.

- Review all reading assignments for this lesson, especially the Chapter 10 and Chapter 11 Summaries on pages 167 and 188 in the Starr textbook (Chapter 11 and Chapter 12 Summaries on pages 189 and 214 respectively in the Starr/Taggart textbook) and the Viewing Notes in this lesson.

- Write brief answers to all the Review Questions at the end of Chapters 10 and 11 of the Starr textbook (Chapters 11 and 12 in the Starr/Taggart textbook) to be certain you understand the text material.

- Complete the Review Activities in this guide to reinforce your understanding of important terms and concepts. Check your answers with the Answer Key and review when necessary.

- Take the Self-Test in this guide to measure your achievement of the Learning Objectives. Check your answers with the Answer Key and review when necessary.

- Complete the Using What You've Learned activities and any other activities and projects assigned by your instructor.

Overview

In Lesson 6, you explored the mechanisms of cellular reproduction and how genes from different parents are transmitted to offspring. In this lesson, you will examine more closely the laws which govern the way in which these genes are outwardly expressed. These are the laws of inheritance.

The laws of inheritance were first postulated by the Austrian monk Gregor Mendel. From his crossbreeding experiments with garden pea plants, he was able to trace the transmission of traits such as flower color. From detailed notes recording the results of thousands of generations of crossbred pea plants, Mendel was able to estimate the mathematical probability of certain traits showing up in successive generations.

Although Mendel's work was largely ignored at the time, it later became the basis for understanding the concepts of dominance and recessiveness. Since that time, the laws of simple inheritance have been expanded to explain a host of related phenomena, as when a parent of one color and a parent of another color produce offspring of a color blending both.

From familiar examples, you will learn how these laws of inheritance affect every aspect of hereditary transmission, from the inheritance of blood type to the inheritance of blood disorders such as sickle-cell anemia. You'll also learn about the effects of environmental factors on inheritance of certain traits.

As you continue your exploration of inheritance, you will learn about sex chromosomes and sex-linked genes. You will learn not only how gender itself is determined, but how gender-related traits are transmitted from one generation to the next. You will survey some of the genetic variations that occur—good and bad—when sex-linked genes and chromosomes are redistributed during meiosis.

Finally, you will learn of rare genetic disorders that can occur when chromosomes undergo structural changes or fail to separate properly during gamete formation. With an awareness of these genetic disorders and their devastating effects, you gain newfound appreciation for medical technologies such as genetic screening and prenatal diagnosis.

Learning Objectives

When you have completed all assignments for this lesson, you should be able to:

1. Use Mendel's experiments to illustrate terms commonly used in genetics and to explain the principles of *dominance*, *segregation*, and *independent assortment*.

2. Determine all the possible kinds of gametes that can be formed from a given *genotype*, and use this information with a Punnett square to predict the outcomes of *monohybrid* and *dihybrid* crosses.

3. Describe variations that occur in observable patterns of inheritance that cannot be explained by simple dominance or recessiveness.

4. State the relationship of genes to chromosomes, and use this information to describe the concept of *linkage* and the likelihood of "crossing over."

5. Distinguish patterns of *autosomal inheritance* from those of *X-linked inheritance*.

6. Explain how changes can occur in chromosome structure and number, and understand how these changes can affect the outward appearance of organisms.

7. Identify and use the components of a simple *pedigree*.

8. Explain how modern methods of genetic screening can minimize potentially tragic events.

Viewing Notes

The video program for Lesson 7 uses a highly visual approach to clarify the laws of inheritance. Beginning with Gregor Mendel's own experiments with plant breeding, you'll see how curiosity, scientific dedication, and painstaking research led Mendel to discover one of the most important secrets of life: how characteristics are passed down from one generation to the next.

From Mendel's experiments, you'll time travel to the present to see how contemporary geneticists such as Kirk Larson breed new forms of fruits and vegetables. In the process, you'll learn something about how plants reproduce naturally and how this reproductive process can be modified by human intervention.

With Dr. Marvin Rosenberg as your guide, you'll learn more about the laws of inheritance and how alleles—different molecular forms of the same gene—have either a dominant or a recessive influence on the inheritance of a particular trait. You'll see how a Punnett square is used to predict the possible allele combinations and the trait likely to be inherited in each case. With the help of Dr. Kevin Moses, you'll learn about some of the exceptions to the laws of inheritance, such as codominance, where both dominant and recessive alleles are expressed independently of each other; incomplete dominance, where the combination of two alleles produces a trait that is a hybrid of dominant and recessive traits; and epistasis, when a particularly concentrated form of an allele masks both dominant and recessive traits. This later phenomenon is dramatically demonstrated in Part Two by Mel Carpenter, an entrepreneur who breeds white pigeons for papal ceremonies and other events.

In Part Three, "Blood Relatives," you'll leave the world of plants and pigeons to explore the role of inheritance in the transmission of human diseases such as hemophilia. You'll learn how a pedigree chart can be used to track the transmission of a particular gene and the trait it expresses. Computer simulation will provide a graphic demonstration of how such a trait is passed from one generation to the next, and Dr. Carol Kasper will provide a medical interpretation of the disease. You'll also meet hemophiliac Brian Craft, who will explain what it's like to live with hemophilia.

As you watch the video program, consider the following questions:

1. How do the experiments of modern plant breeders differ from the methods used by Gregor Mendel?
2. How was Mendel able to isolate a particular trait for study? How did he control for outside factors that might confound his results?
3. What is the reproductive mechanism that allows for an offspring to inherit a trait that is expressed in neither parent?
4. How does a Punnett square work?
5. What is the theory of independent assortment?
6. What are some of the phenotypic variations that are possible with codominance? With incomplete dominance?

7. What is epistatis, and how does it influence the inheritance of a particular trait?
8. What is hemophilia? What kind of inherited defect produces the disorder?
9. What are some of the signs and symptoms of hemophilia?
10. How does the process of inheritance in humans differ from that of plants?

Review Activities

Matching

Match the terms listed below with the definitions that follow. Check your answers with the Answer Key and review any terms you missed.

I.

____ 1. allele
____ 2. autosome
____ 3. genotype
____ 4. homologous chromosome
____ 5. karyotype
____ 6. pedigree
____ 7. phenotype
____ 8. sex chromosome
____ 9. X-chromosome
____ 10. Y-chromosome

   a. a standardized chart showing the genetic connections among related individuals

   b. a chromosome that is of the same number and kind in both males and females

   c. the chromosome which usually determines the offspring's gender

   d. in species that reproduce sexually, one of a pair of chromosomes that resemble each other in size, shape, and the genes they carry

   e. the chromosome in human beings that carries the genetic code to produce the male of the species

   f. the genetic composition of an organism

   g. the observable trait or traits of an individual that arise from genetic interactions

   h. one of two or more molecular forms of a gene that produces different versions of the same trait

   i. the chromosome in human beings that carries the genetic code to produce the female of the species, provided that two are inherited

   j. the number of chromosomes, or sum of the characteristics of the chromosomes, of a eukaryotic cell

II.

___ 1. amniocentesis   ___ 6. inversion
___ 2. aneuploidy   ___ 7. nondisjunction
___ 3. crossing over   ___ 8. polyploidy
___ 4. deletion   ___ 9. translocation
___ 5. epistasis

a. the process whereby a fluid sample is drawn from a fetal sac of a pregnant female

b. the process during meiosis in which non-sister chromatids of homologous chromosomes break at the same place along their length and exchange corresponding segments

c. a change in the number of chromosomes following inheritance of one less chromosome or one more chromosome

d. a change in chromosome structure after a region is lost or damaged

e. the process whereby two alleles of one gene influence the expression of alleles of a different gene

f. a change in chromosome structure following the insertion of part of a nonhomologous chromosome into it

g. the failure of one or more chromosomes to separate during cell division

h. a change in chromosome structure that occurs after a segment separated from the chromosome, then inserted at the same place in reverse

i. a change in chromosome number caused by the inheritance of three or more of each type of chromosome

III.

___ 1. codominance
___ 2. continuous variation
___ 3. dihybrid cross
___ 4. heterozygous
___ 5. homozygous
___ 6. incomplete dominance
___ 7. independent assortment
___ 8. in vitro fertilization
___ 9. monohybrid cross
___ 10. segregation

a. when a trait appears that is somewhere between the homozygous dominant and homozygous recessive version of the trait

b. a condition in which nonidentical alleles are both expressed even though they code for two different phenotypes

c. characterized by having nonidentical alleles at a particular location on a homologous chromosome

d. the principle that each gene pair tends to wind up in gametes independently of other gene pairs on nonhomologous chromosomes

e. the process of fertilizing a female egg outside of the mother's body and in an artificial environment

f. crossbreeding two individuals who differ in two traits rather than one

g. characterized by having identical alleles at a particular location on a homologous chromosome

h. the principle that gene pairs separate during meiosis and end up in different gametes

i. crossbreeding which ensures that each offspring inherits a pair of nonidentical alleles for a single trait, and is therefore heterozygous

j. a range of minor differences in a given trait among all individuals in a population.

Completion

Fill in each blank with the most appropriate term from the list for that paragraph. A term may be used once, more than once, or not at all. If a question requires two or more answers in succession, they may be in any order (unless the question indicates otherwise). Check your answers with the Answer Key and review when necessary.

1. Through his experiments with _____ crosses between true-breeding plants, Mendel was able to deduce that _____ organisms have two _____ for each trait, and that these _____ are transmitted intact to offspring. He found that the trait expressed in the offspring depended on which gene was _____. His work led to the laws of _____ including the law of _____, which explains how pairs of genes on a set of homologous chromosomes wind up in different _____. His experiments with crosses led to the theory of _____, which explains how gene pairs on two sets of homologous chromosomes are sorted for distribution.

dihybrid	monohybrid
diploid	monoploid
dominant	one
gametes	recessive
gene(s)	segregation
independent assortment	two
inheritance	

2. Modern geneticists use a variety of tools to carry on the work of Mendel. _____ are still an effective means of predicting the outcomes of simple inheritance, however, inheritance is often complicated by _____ and _____. These complications often result in _____, where the offspring exhibits a trait that represents a blend of parental traits, and _____, where two traits are expressed at the same time by a pair of alleles. _____ describes a phenomenon where single genes have an effect on multiple traits, while _____ describes a situation where one gene pair influences the phenotypic expression of another gene pair.

codominance	independent assortment
continuous variation	pleiotropy
environmental factors	Punnett squares
epistasis	segregation
gene interactions	testcrosses
incomplete dominance	

84 Lesson 7 / Patterns of Inheritance

3. The transmission of sex chromosomes and _____ genes also influences the kinds of traits offspring will inherit. Genetic disorders such as _____ occur when the male inherits a _____, _____ allele. Through the mechanism of _____ the genetic deck can be shuffled even further. _____ occurs when segments of _____ genes are swapped between non-sister chromatids during _____. In rare cases, chromosomes fail to separate properly (_____, leading to genetic disorders such as _____ (3 chromosome 21), _____ (1 X chromosome only), and _____ (2 X and 1 Y chromosome). Other disorders may be caused by structural changes to the chromosome including _____, _____, _____, and _____.

crossing over	meiosis
deletion	mitosis
duplication	nondisjunction
dominant	recessive
Down syndrome	sex-linked
inversion	translocation
Klinefelter syndrome	Turner syndrome
linked	X-linked
hemophilia	Y-linked

Self-Test

Select the one best answer for each question. Check your answers with the Answer Key and review when necessary.

1. When Mendel crossbred purple-flowering plants and white-flowering plants, he discovered that

    a. one cannot predict flower color from one generation to the next.

    b. purple flowers are a dominant trait.

    c. the alleles for flower color and size are on the same chromosome.

    d. the offspring had lighter purple or lavender flowers.

2. Mendel's experiments led to the proposition that pairs of homologous chromosomes (and the genes they carry) are sorted out and packaged in gametes, regardless of how other pairs are sorted out. This proposition is now known as the theory of

    a. continuous variation.

    b. gamete-driven chromosome pairing.

    c. independent assortment.

    d. genetic individuation.

3. If B is a dominant allele and allele b is its recessive counterpart, which genotypes can be formed through a monohybrid cross of one organism with Bb and another with bb?

    a. Bb, bb

    b. BB, bb

    c. BB, Bb

    d. BB, Bb, bb

4. When crossbreeding some species of flowers, the color of the offspring flowers may be somewhere between the colors of the parent flowers. This phenomenon is an illustration of

    a. a multiple allele system.

    b. incomplete dominance.

    c. pleiotropy.

    d. epistasis.

5. Genes that are "linked" to a particular chromosome
   a. produce the same trait in every generation.
   b. are the only genes associated with a specific site on the chromosome.
   c. cannot cross over to the homologous chromosome during mitosis.
   d. end up in the same gamete.

6. Sex-linked inheritance differs from autosomal inheritance in that sex-linked inheritance
   a. determines only the gender of the offspring.
   b. involves gamete formation and cross-fertilization.
   c. produces traits that are found in both males and females.
   d. produces traits that are found more frequently in either males or females, but not both.

7. Individuals with Down syndrome, Turner syndrome, and Klinefelter syndrome are similar in that
   a. they have a duplicated gene sequence on one of their chromosomes.
   b. they have lost a region of a chromosome and the genes linked to it.
   c. they have an abnormal number of chromosomes.
   d. a section of one chromosome has translocated to another, nonhomologous chromosome.

8. Which of the following mechanisms best explains the syndromes listed above?
   a. abnormal duplication of a single gene
   b. nondisjunction or meiosis
   c. during fertilization, one gamete has an extra chromosome
   d. b and c

9. In the following pedigree diagram, the symbol ● represents

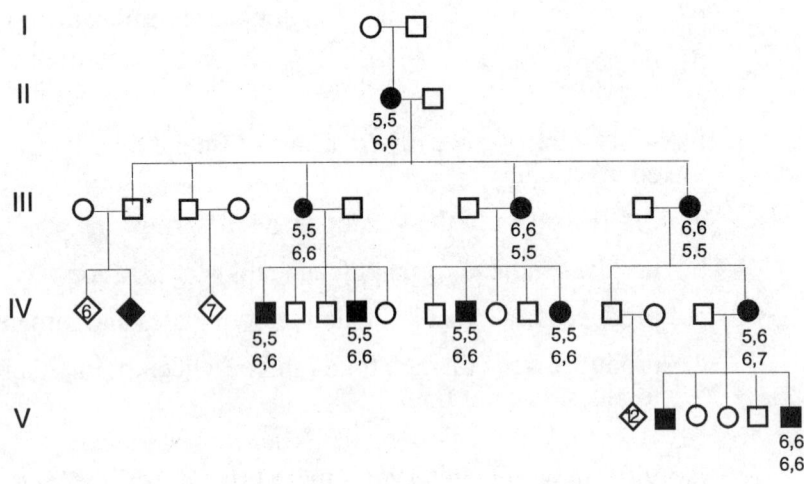

* Gene not expressed in this carrier

   a. any female.
   b. an individual whose trait is being tracked.
   c. male offspring.
   d. a genetic abnormality.

10. Genetic screening is a technique used to

   a. remove defective genes from an individual chromosome.
   b. detect genetic abnormalities in unborn fetuses.
   c. selectively breed plants and animals that have desirable traits.
   d. identify and treat persons who carry genes for inherited disorders.

Using What You've Learned

Based on your own interests or your instructor's requirements, complete one or more of the following activities.

1. Toss a coin ten times and record the number of heads and tails. Repeat the experiment tossing the coin one hundred times. Explain the importance of large sample size and variation from expected ratios.

2. Create a pedigree chart. Select a trait and survey your family members for presence or absence of the trait.

3. List several reasons why Mendel's pea plants were a natural choice for a genetics experiment. Name an experimental organism that would be a poor choice for genetic research; explain your answer.

4. The ABO system of blood typing includes four possible types: A, B, AB, and O. Identify the different alleles for this trait. (Hint: there are three.)

5. Contact your local hospital or doctor's office to request brochures explaining various genetic problems that can be inherited. Which problems require mandatory prenatal testing in your state? Which allow the option of voluntary testing?

6. Learn more about hemophilia. Research its history, its involvement in the downfall of the Russian monarchy, and breakthroughs in modern treatments for the disease.

7. Research the concept of genetic screening. What kinds of technologies and procedures are used? How effective are these technologies and procedures?

Challenge Questions

1. If monohybrid crosses produce four possible genotypes, and dihybrid crosses produce 16 possible genotypes, how many genotypes would be produced by a trihybrid cross (three sets of alleles)?

2. If recessive genes are transmitted from one generation to the next, how do you explain the disappearance of certain traits from entire populations?

3. What kinds of traits provide good examples of simple dominance? Which kinds of traits are not good examples?

4. What is the difference between codominance and incomplete dominance in terms of how the inherited trait is expressed in offspring?

5. How many alleles does it take to produce the four possible blood types found in human beings?

6. How can blood type be used to establish paternity?

7. How do researchers know whether a new strain of organism represents a new species or a phenotypic variation of an existing species?

LESSON 8

DNA: Structure and Function

Assignments

For the most effective study of this lesson, we suggest that you complete the following assignments in the sequence listed below:

Before Viewing the Video Program
- Read the Overview and Learning Objectives for this lesson. Use the Learning Objectives to guide your reading, viewing, and thinking.

- Read Chapter 12, "DNA Structure and Function," pages 190–199, and Chapter 15, "Recombinant DNA and Genetic Engineering," pages 222–236, in the Starr textbook.

    Or read, in the Starr/Taggart textbook, Chapter 13, pages 216–225, and Chapter 16, pages 256–268.

- Read the Viewing Notes in this lesson.

View the "DNA: Blueprint of Life" Video Program
After Viewing the Video Program
- Briefly note your answers to the questions at the end of the Viewing Notes.

- Review all reading assignments for this lesson, especially the Chapter 12 and Chapter 15 Summaries on pages 199 and 234–235 in the Starr textbook (Chapter 13 and Chapter 16 Summaries on pages 225 and 267 respectively in the Starr/Taggart textbook) and the Viewing Notes in this lesson.

- Write brief answers to all of the Review Questions at the end of Chapters 12 and 15 in the Starr textbook (Chapter 16 in the Starr/Taggart textbook) to be certain you understand the text material.

- Complete the Review Activities in this lesson to reinforce your understanding of important terms and concepts. Check your answers with the Answer Key and review when necessary.

- Take the Self-Test in this lesson to measure your achievement of the Learning Objectives. Check your answers with the Answer Key and review when necessary.

- Complete the Using What You've Learned activities and any other activities and projects assigned by your instructor.

Overview

Until now, you have studied the mechanisms of cellular division and the laws of genetic inheritance. At the center of these processes are the extraordinary chromosomes, which contain the entire blueprint for our existence. While the chemical composition of the chromosomes remained a mystery until recently, chromosomes' impact on the survival and perpetuation of all living things depends on these important molecules of life.

In Lesson 8, you will get your first look at the chemical nature of the chromosome, which is composed primarily of deoxyribonucleic acid, commonly known as *DNA*. You will learn about the various experiments that confirmed the existence of DNA and eventually established its chemical composition. You will follow developments that led to the historic discovery of the double-stranded helical structure that provided the theoretical basis for molecular biology and the model for today's genetic engineers.

While the search for the composition and structure of DNA is an intriguing one, you will be fascinated by the manner in which our hereditary blueprint is encoded on the DNA molecule. Four nitrogen-based nucleotides joined in sequences provide all the chemical variability needed for the incredible number of genetic possibilities found in all living things. In the process of exploring DNA's basic chemical structure, you will learn how the DNA molecule repairs and replicates itself prior to cellular division.

In addition to providing a foundation for understanding the biochemistry of reproduction and the laws of heredity, your knowledge of DNA will give you new insights into the rapidly growing field of genetic engineering. The lesson will introduce you to some of the more recent discoveries in recombinant DNA technology and how this technology is being put to use. You will be asked to consider the ethics of manipulating genetic material to eliminate hereditary diseases or to create new life forms and to speculate on what the future will bring as the techniques of genetic engineering are expanded and improved.

Learning Objectives

When you have completed all assignments for this lesson, you should be able to:

1. Relate how past experiments demonstrated that instructions for producing inheritable traits are coded in DNA.

2. Describe the parts of a *nucleotide*, and explain how nucleotides are linked together to make DNA.

3. Explain how DNA is replicated and repaired, what materials are needed for replication, and the importance of that process to cell division.

4. Outline the mechanism through which DNA controls a cell's structure and function.

5. Explain what *plasmids* are and how they may be used to make *recombinant DNA* molecules.

6. Describe in general terms how DNA can be cleaved, spliced, cloned, and sequenced.

7. Discuss some limits and possible applications of genetic engineering, and distinguish this process from the more traditional techniques of selective breeding.

8. Describe some of the methods scientists use to study DNA.

Viewing Notes

In the video segment for Lesson 8, "DNA: Blueprint of Life," the secrets of the genetic code are revealed along with their implications for medical science. Part One reviews the historical developments that led to the isolation and eventual identification of the hereditary material we call DNA. You'll hear geneticists Alfred Hershey and James Watson describe their respective contributions to the landmark discovery of the double-stranded helical structure that characterizes DNA, a discovery at the heart of modern molecular biology.

With the guidance of Dr. Leroy Hood and the three-dimensional perspectives provided by computer simulation, you'll explore the structure of DNA, including nitrogen-containing bases that form the "steps" of this molecular staircase. You will see how a limited number of nucleotides can produce seemingly limitless genetic variability simply by the way these nucelotides are sequenced in the chromosome.

The medical implications of DNA structure and replication are raised in Part Two, "Good Genes, Bad Genes." Dr. Donald Kohn describes the case of Andrew Gobea, who inherited a genetic illness known as Severe Combined Immune Deficiency or SCID. The disease, which interferes with the production of white blood cells and the overall effectiveness of the immune system, results when a crucial fragment of DNA is damaged or missing. Using the relatively new techniques of gene therapy, Dr. Kohn explains how Andrew's defective DNA was identified and reintroduced in an attempt to establish normal genetic content.

In Part Three, "Genetic Detectives," Dr. John Wasmuth and Dr. Inder Verma describe what may be the most ambitious scientific project ever undertaken: the human genome project, the identification of each and every one of the possibly 100,000 human genes. Professors Wasmuth and Verma will introduce you to some of the techniques used to isolate the genes for the treatment of specific ailments and the progress that has been made to date. By the conclusion of the program, you will wonder, as they do, what the future holds as we begin to harness the awesome power of these discoveries.

As you watch the video program, consider the following questions:

1. What scientific efforts ultimately led to the discovery of the molecular structure of DNA by Watson and Crick?
2. What are the chemical components of DNA, and how do these components combine to form a DNA molecule?
3. How are the instructions for various hereditary traits encoded using only four nucleotides?
4. How does DNA replicate itself?
5. How can recombinant DNA techniques be used to treat genetic defects and restore normal function in a given organism?
6. How is recombinant DNA "amplified" and delivered to new host cells?
7. How might medical practice be affected by the refinement of gene therapy?

8. What techniques are being used to map specific genes on the DNA molecule?

9. What are the benefits and risks associated with increased use of genetic engineering?

Review Activities

Matching

Match the terms listed below with the definitions that follow. Check your answers with the Answer Key and review any terms you missed.

I.

___ 1. DNA
___ 2. DNA ligase
___ 3. DNA polymerase
___ 4. gene
___ 5. nucleotide
___ 6. plasmid
___ 7. adenine
___ 8. cytosine

a. in bacteria, a small circular molecule of extra DNA that carries only a few genes

b. the unit of information about an inheritable trait that is passed on from parents to offspring

c. an enzyme that assembles a new strand of DNA on a parent DNA strand during replication

d. a small organic compound that is the basic structural unit of DNA

e. the common name for deoxyribonucleic acid

f. an enzyme that joins together nucleotides during DNA replication and repair

g. a nucleotide that base pairs with thymine

h. a nucleotide with a single-ring structure

II.

___ 1. cloning   ___ 5. genetic engineering

___ 2. gene sequencing   ___ 6. recombinant DNA technology

___ 3. gene splicing   ___ 7. reverse transcription

___ 4. gene therapy

a. the process of transferring normal genes into the body cells of an organism to correct genetic defects

b. the process of assembling DNA on a single-stranded mRNA molecule by viral enzymes

c. the process of determining the arrangement of base pairs on a given DNA fragment

d. the process of producing multiple, identical copies of DNA fragments that have been inserted into a vector

e. the process of altering the information content of DNA through the use of various recombinant techniques

f. the process of inserting DNA fragments from an outside source into the DNA molecule of a host cell

g. the processes by which DNA from different species may be cleaved, spliced together, and regenerated in quantity

Completion

Fill in each blank with the most appropriate term from the list for that paragraph. A term may be used once, more than once, or not at all. If a question requires two or more answers in succession, they may be in any order (unless the question indicates otherwise). Check your answers with the Answer Key and review when necessary.

1. The hereditary material we now call _____ was first isolated by _____ in 1869. In 1928, _____ discovered that hereditary material could be transferred from one _____ cell to another, causing a permanent change in the host cell and its descendants. Twenty years later, _____ and _____ used _____ to isolate and identify the hereditary material as _____. Drawing from _____ and other data, _____ and _____ finally proposed a three-dimensional structural model for DNA in 1953. This structure, the _____, became the basis for understanding DNA replication and repair.

   animal
   bacterial
   Chase
   Crick
   DNA
   double-stranded helix
   electron microscopy
   Franklin
   Griffith
   Hershey

   Miescher
   nucleic acid hybridization
   Pauling
   plant
   radioactive isotopes
   single-stranded coil
   X-ray crystallography
   Watson
   triple-coiled ladder

2. DNA contains four types of _____. Each nucleotide consists of a _____-carbon sugar, a(n) _____ group, and one of the following nitrogen-containing bases: _____ (A), _____ (G), _____ (T), and _____ (C). While only _____ and _____ base pairs are possible, the lengthy sequence of base pairs along the _____ provides endless hereditary possibilities. During DNA replication and repair, the _____ bonds holding the base pairs together are broken by _____. DNA _____ then attach free _____ to the exposed bases, and DNA _____ seal the new _____ segments into a continuous strand.

A-C	G-T
A-G	glycine
A-T	guanine
adenine	histone
amino	hydrogen
C-G	ligases
C-T	nucleosome
chromosome	nucleotide(s)
cytosine	phosphate
cysteine	polymerases
enzymes	sulfide
five	thymine
four	trypsin

3. Today, scientists are using their growing knowledge of DNA structure and function in _____. Using _____, they have been able to isolate, modify, and reintroduce _____ into _____ for the purpose of correcting genetic defects or introducing new features into plant and animal populations. _____ uses specific _____ to cut the _____ at specific points. The resulting _____ are then inserted into _____ for _____. _____ such as _____ are often used to deliver the modified DNA to the _____.

base pairs	nucleosomes
cloning	nucleotides
DNA strands	plasmids
DNA fragments	recombinant DNA technology
genes	restriction enzymes
host cells	reverse transcription
genetic engineering	sequencing
histones	vectors
nucleic acid hybridization	viruses

Self-Test

Select the one best answer for each question. Check your answers with the Answer Key and review when necessary.

1. While attempting to create a vaccine for *Streptococcus pneumoniae*, Griffith (1928)

   a. discovered the process by which DNA replicates itself prior to cellular division.

   b. pioneered the use of recombinant DNA technology using bacterial cells.

   c. identified the chemical composition of DNA from nuclear material extracted from laboratory mice.

   d. demonstrated that the genetic instructions for infection could be passed from one generation of cells to the next.

2. The four nucleotides that make up DNA are similar in that

   a. they contain a five-carbon sugar and nitrogen-containing base.

   b. they have a purine (double-ring) molecular structure.

   c. they are connected to each other by way of phosphate bonds.

   d. the base of each nucleotide can pair up with any other base.

3. The natural structure of the DNA molecule is based on a

   a. single, tightly coiled strand.

   b. flat, ladder-like shape, with nucleotides as the rungs.

   c. double-stranded helix.

   d. central strand with unpaired bases flaring outward.

4. The replication of DNA depends on

   a. the breakdown of DNA strands into free nucleotides.

   b. the breakdown and resealing of nucleotides on the DNA strands through enzymes.

   c. adding new nucleotide segments to intact DNA strands.

   d. the chemical transformation of guanine to cytosine.

5. Plasmids are important in recombinant DNA technology because they

   a. are the machinery for cloning DNA fragments.

   b. carry only one type of gene that can be easily identified.

   c. manufacture the restriction enzymes needed for cutting DNA.

   d. are the basis for viruses that can deliver genes to a new host.

6. Splicing new DNA segments into a host cell's chromosomes depends on all of the following EXCEPT

   a. the availability of a vector that can deliver the DNA to a target host cell.

   b. the ability of restriction enzymes to cut specific nucleotide sequences.

   c. the chemical modification of nucleotides to create new base pair combinations.

   d. the exposure of single-stranded "sticky ends" that can base pair with other DNA fragments.

7. Genetic engineering has already been used to

   a. develop a cure for Alzheimer's disease.

   b. produce strains of bacteria to clean up oil spills.

   c. create species of plants that can thrive in low-gravity environments.

   d. determine the sex of unborn infants.

8. The human genome project is an international effort to

   a. clone human beings to provide organs for transplant operations.

   b. decipher the total genetic information carried by the human organism.

   c. eliminate all inheritable diseases from the human gene pool.

   d. produce a complete DNA molecule through artificial means.

Using What You've Learned

Based on your own interests or your instructor's requirements, complete one or more of the following activities.

1. Using a set of labeled paper shapes representing the sugars, phosphate groups, and the four nitrogen bases present in DNA, construct a short segment of a DNA strand. If possible, have a colleague construct the complimentary strand according to the rules of base pairing. Use the finished construction to show the semiconservative replication of DNA.

2. Prepare a chronological listing of significant contributions to the discovery of DNA structure. Describe the principal methods used.

3. Select a research tool or technique that is currently used to study DNA structure and function. Find additional information on the tool and prepare a report summarizing how it works and its contribution to science.

4. Prepare a chronological listing of significant contributions to the field of genetic engineering.

5. Identify a genetically engineered product that is currently in use. Research the development of this product and how it benefits society. Also, identify any potential risks to society. Prepare a report summarizing your findings.

6. Prepare a table summarizing the DNA recombination methods discussed in Chapter 13. Use illustrations where needed to clarify each method.

7. Prepare an argument *for* or *against* genetic engineering use and development. If possible, debate your position with a colleague who supports the opposing viewpoint.

8. Using the set of labeled paper shapes introduced in item 1, demonstrate how DNA is cleaved, cloned, and spliced using recombinant DNA methods.

Challenge Questions

1. Why do you think DNA evolved as a double-stranded molecule rather than a single-stranded molecule?
2. What is semiconservative replication and what are its advantages?
3. How do you explain the fact that the DNA of all single-celled and multicellular organisms uses the same nucleic acids?
4. What happens if DNA is damaged? How does the organism recognize and repair such damage?
5. Does genetic recombination naturally occur? If so, where and when does it occur?
6. What kinds of organisms are ideally suited for genetic engineering research and why?
7. What kinds of problems have arisen as a result of widespread use of antibiotics, and how may they be explained in terms of DNA structure and function?
8. In what ways might genetically engineered organisms pose a threat to the health and safety of human populations? What kinds of safeguards exist to reduce this threat?

LESSON
9

Proteins

Assignments

For the most effective study of this lesson, we suggest that you complete the assignments in the sequence listed below:

Before Viewing the Video Program

- Read the Overview and Learning Objectives for this lesson. Use the Learning Objectives to guide your reading, viewing, and thinking.

- Read Chapter 13, "From DNA to Proteins," pages 200–211, and Chapter 14, "Controls Over Genes," pages 212–221, in the Starr textbook.

  Or read, in the Starr/Taggart textbook, Chapter 14, pages 226–241, and Chapter 15, pages 242–255.

- Read the Viewing Notes in this lesson.

View the "Proteins: Building Blocks of Life" Video Program
After Viewing the Video Program

- Briefly note your answers to the questions at the end of the Viewing Notes.

- Review all reading assignments for this lesson, especially the Chapter 13 and Chapter 14 Summaries on pages 210 and 220 in the Starr textbook (Chapter 14 and Chapter 15 Summaries on pages 238 and 254 respectively in the Starr/Taggart textbook) and the Viewing Notes in this lesson.

- Write brief answers to all the Review Questions at the end of Chapters 13 and 14 in the Starr textbook (Chapters 14 and 15 in the Starr/Taggart textbook) to be certain you understand the text material.

- Complete the Review Activities in this guide to reinforce your understanding of important terms and concepts. Check your answers with the Answer Key and review when necessary.

- Take the Self-Test in this guide to measure your achievement of the Learning Objectives. Check your answers with the Answer Key and review when necessary.

- Complete the Using What You've Learned activities and any other activities and projects assigned by your instructor.

Overview

In previous lessons, you were introduced to the mechanisms of cellular reproduction and heredity. You have learned, for example, how the genetic instructions from parents are packaged and transmitted to offspring, and what kinds of physical traits are likely to result from different chromosome configurations.

How does the genetic code translate into eye color or plant height? The answer is proteins. The genetic code is essentially a manual for the production of proteins, for proteins control cellular metabolism, differentiation, and the expression of different phenotypic traits in the organism. In this lesson, you will discover how the genetic code is "transcripted" from a master DNA molecule to a modified version of DNA called RNA and how the code is then "translated" into the amino acid sequences that characterize different proteins.

As usual, your exploration of these concepts will involve some reading. The technical explanations provided in your textbook will be further illustrated by the examples and animations in the videotape accompanying the lesson. Together, these two resources will provide you with the necessary background to understand some of the most interesting—and occasionally deadly—phenomena in nature.

These phenomena range from the mysterious slime mold, a plantlike organism which is able to change its form and move across the forest floor, to *E. coli* bacteria, to the tumor cells which signal the onset of cancer in human beings. As you'll soon discover, these phenomena have more in common than you may think.

While many of the concepts presented in this lesson will seem abstract and theoretical at first, you will begin to see how these concepts play a role in the inheritance of genetic disorders such as sickle-cell anemia. You will also see how science is progressing in its quest to unravel the mysteries of genetic code and how the knowledge gained thus far is providing new hope for curing such disorders.

Learning Objectives

When you have completed all assignments for this lesson, you should be able to:

1. State the major differences between DNA and RNA, and describe how the structure of DNA determines the structure of the three forms of RNA during *transcription*.

2. Explain how the structure and function of the three forms of RNA determine the primary structure of polypeptide chains during *translation*.

3. Outline how *operons* regulate gene expression (the production of proteins) in prokaryotes.

4. Describe how cell differentiation proceeds in eukaryotes by selective gene expression during development.

5. Relate the concept of changes in gene controls to changes in protein. Note those changes that occur naturally, as in some cancers, and those that can occur through genetic engineering.

6. Describe some of the experimental methods scientists use to investigate protein transcription and translation.

Viewing Notes

In "Proteins: Building Blocks of Life," many of the concepts you read about in your text will be vividly illustrated through computer-animated sequences and real-life case studies. In Part One, for example, Dr. James Lake introduces the how and why of RNA transcription and translation. His introduction is followed by a demonstration of the process at the molecular level. You'll watch as enzymes unwind the DNA molecule, making room for the nucleotides that will eventually become mRNA, tRNA, and rRNA. You'll also see how the components for protein synthesis converge on the ribosome and assemble the amino acids into three-dimensional polypeptide chains.

Dr. Lake points out recent discoveries suggesting that prokaryotic and eukaryotic cells function very similarly at the level of protein synthesis. Yet even though the process seems to work well for bacteria and humans alike, it is not without its shortcomings. In Part Two, you'll learn what happens "When Proteins Go Bad" due to minor alterations of the genetic code. You'll meet weatherman Christopher Nance, who was diagnosed with sickle-cell anemia as a child and given only a few years to live. You'll also hear from Dr. Cage Johnson, who will describe the mechanism that produces the sickle-shaped blood cells characteristic of the disease. In the process, you'll gain an appreciation for the profound pathological consequences that can result when a minor adjustment in the genetic code—in this case, the substitution of the nucleotide base adenine for thymine—disrupts protein synthesis.

Part Three discusses another potentially lethal disorder, cancer, and its relationship with a most interesting organism called the slime mold. Through the study of slime molds, scientists such as Dr. Richard Firtel are discovering the mechanisms that control cell differentiation and the selective expression or suppression of genes. The more science learns about these mechanisms, the closer we come to understanding and treating disorders such as cancer, which occurs when normal controls over cell growth and differentiation go awry.

As you watch the video program, consider the following questions:

1. What does Dr. Lake mean when he refers to proteins as the "bricks and mortar of the house"?
2. What are the different types of RNA and what are their functions?
3. How are amino acids linked together to form three-dimensional polypeptide chains?
4. How does the study of bacteria cells help us understand how protein synthesis works in human cells?
5. How does sickle-cell anemia illustrate the consequences of errors in the genetic code?
6. Why is sickle-cell anemia considered a life-threatening disease? How does it affect the health of an individual?
7. How does slime mold help scientists understand how genetic control systems work?

8. How does cell differentiation occur? Why are certain genes expressed and others not expressed in a given cell?

9. What is the mechanism that causes cancer, and what are its causes?

Review Activities

Matching

Match the terms listed below with the definitions that follow. Check your answers with the Answer Key and review any terms you missed.

I.

___ 1. activator protein

___ 2. anticodon

___ 3. base sequence

___ 4. codon

___ 5. mRNA

___ 6. RNA

___ 7. rRNA

___ 8. transcription

___ 9. translation

___ 10. tRNA

a. a regulator protein that plays a major role in metabolic control systems

b. the conversion of the coded information in messenger RNA into a specific sequence of amino acids

c. ribonucleic acid, a single-stranded nucleic acid

d. one of a series of base triplets in a messenger RNA molecule

e. the order in which each nucleotide base follows the next in a strand of DNA or RNA

f. a sequence of three nucleotide bases in a transfer RNA molecule that can pair with a messenger RNA codon

g. a type of RNA molecule that combines with proteins to form ribosomes

h. a type of RNA molecule that binds and delivers specific amino acids to ribosomes

i. the assembly of an RNA strand on one of the two strands of a DNA molecule

j. a series of ribonucleotides transcribed from DNA and translated into a polypeptide chain

II.

___ 1. Barr body ___ 5. metastasis
___ 2. cancer ___ 6. mutagen
___ 3. carcinogen ___ 7. oncogene
___ 4. cell differentiation

a. a malignant tumor consisting of cells characterized by abnormal growth and division

b. an environmental substance that can permanently modify the structure of a DNA molecule

c. a gene having the potential to trigger cancerous changes in a cell

d. the process by which a cell suppresses part of its genes in order to become specialized in composition, structure, and function

e. an environmental substance that can trigger cancer

f. the transfer of cancerous cells from the site of origin to another part of the body

g. a condensed X chromosome that was deactivated during early embryonic development

Completion

Fill in each blank with the most appropriate term from the list for that paragraph. A term may be used once, more than once, or not at all. If a question requires two or more answers in succession, they may be in any order (unless the question indicates otherwise). Check your answers with the Answer Key and review when necessary.

1. The instructions for building proteins are contained in the _____ sequences of _____. The process begins when a double-stranded _____ molecule is unwound, exposing the _____. In a process called _____, enzymes called _____ assemble _____ on these exposed bases. The only difference between this process and DNA replication is that _____ pairs with _____ rather than thymine. Through the process of _____, RNA directs the linkage of one amino acid to another to produce the _____ chains that make up a protein.

adenine	nucleotide
amino acids	polymerases
base pairs	polypeptide
DNA	RNA
duplication	termination
elongation	transcription
initiation	translation
guanine	uracil

2. The instructions for protein synthesis are encoded in base _____ called _____. Although the genetic code contains _____ codons, more than one codon may represent a given _____. The sequence of these _____ determines the sequence of _____, and therefore the type of protein being synthesized. In addition to _____, which contains the codons for protein synthesis, two other variations of RNA play important roles in the process. _____ produces the subunits that will anchor _____ to the ribosome, while _____ provides the amino acids and _____ that bind with _____. _____ begins when a small ribosomal subunit, an initiator _____, a _____ transcript, and then a large ribosomal subunit converge.

20	pairs
64	rRNA
amino acid(s)	transcription
anticodon(s)	translation
codon(s)	tRNA
mRNA	triplets

110 Lesson 9 / Proteins

3. _____ occur when nucleotide bases are added, lost, or switched with other bases. While some _____ are spontaneous, others are caused by environmental _____. The substitution of a _____ can result in replacement of an amino acid and synthesis of a defective _____ ,which can lead to genetic disorders such as _____. Depending on the chemical needs of the cell, disruption of normal _____ synthesis can also be deliberate. _____ control systems use regulatory proteins called _____ to block transcription and _____ production of certain proteins. _____ control systems use other regulatory proteins called _____ to _____ production of certain proteins when needed. When the control mechanisms over growth and division break down, _____ may form. _____ tumors are commonly known as _____.

base pair(s)	positive
benign	promoters
cancer	protein(s)
inhibit	repressors
malignant	sickle-cell anemia
mutagens	speed up
mutations	tumors
negative	

Lesson 9 / Proteins   111

Self-Test

Select the one best answer for each question. Check your answers with the Answer Key and review when necessary.

1. How is RNA structurally different from DNA?
   a. The nucleotides in RNA are completely different from those in DNA.
   b. RNA contains only three bases, rather than four, as in DNA.
   c. RNA is seldom single-stranded; DNA is always double-stranded.
   d. One of the four bases in RNA is different from that in DNA.

2. Which of the following carries the amino acids from which polypeptides are synthesized?
   a. mRNA
   b. pRNA
   c. rRNA
   d. tRNA

3. Translation of the genetic code into polypeptide chains begins with
   a. the binding of a promoter with RNA polymerase.
   b. assembly of an RNA strand from a sequence of DNA nucleotides.
   c. the convergence of an initiator tRNA and mRNA transcript on a ribosome.
   d. the attachment of tRNA anticodons to mRNA codons.

4. The sequence of amino acids in a given protein is determined by
   a. the kinds of amino acids bound to tRNA.
   b. the sequence of base triplets or codons on mRNA.
   c. the location where translation takes place on the ribosome.
   d. RNA-guided rearrangement of existing amino acid sequences.

5. In prokaryotic cells, operons
   a. provide a binding site for the activation or repression of subsequent genes.
   b. accelerate the growth and development process that leads to cancer.
   c. bring mRNA and tRNA together prior to polypeptide synthesis.
   d. perform all the functions of mRNA, tRNA, and rRNA as needed.

6. An operon consists of
   a. a sequence of genes.
   b. a promoter followed by a sequence of genes.
   c. an operator followed by a sequence of genes.
   d. a promoter and an operator followed by a sequence of genes.

7. In eukaryotic cells, cell differentiation depends on
   a. controlling the transcription, translation, and expression of genes.
   b. the presence of different genes in different cells.
   c. changing chemical conditions in the cell's internal environment.
   d. the mutation of specific DNA sequences during meiosis.

8. Cancerous tumors are formed when
   a. the chemical composition of an otherwise healthy cell changes.
   b. normal cell DNA is replaced with foreign DNA.
   c. normal controls over cell growth and division are disrupted.
   d. a cell's genetic code is scrambled by environmental mutagens.

9. The effects of gene controls on mRNA transcripts have been verified by observing
   a. the response of plant seedlings to different kinds of light.
   b. the phenotypic expression of Y chromosome inactivation.
   c. the transformation of slime molds.
   d. the inheritance of genetic disorders such as sickle-cell anemia.

10. One of the methods scientists use to study the transcription and translation of the genetic code is
    a. the use of electron microscopes to observe protein synthesis.
    b. developing computer models of proteins based on specific DNA sequences.
    c. tracing small amounts of radioactive DNA introduced into bacteria cells.
    d. observing and analyzing the behavior of simple organisms such as viruses.

Using What You've Learned

Based on your own interests or your instructor's requirements, complete one or more of the following activities.

1. Using a matrix such as one presented in Figure 13.7 (Starr text) or Figure 14.11 (Starr-Taggart text), create several base triplets on 3 x 5 cards. Create two more sets of these cards and distribute each set of cards and a copy of the chart to two other individuals or groups. Have a competition to see which individual/group is first to assemble cards into a given peptide sequence.

2. Using illustrations or three-dimensional materials, show the different ways the instructions in the genetic code can be altered. Which of these changes is/are most important for the evolution of humans and other species?

3. Using illustrations or three-dimensional materials, show the structural differences among the three types of RNA. Show which of the RNAs is reusable and how its structure supports this capability.

4. Visit the library and scan recent science publications for current research on cancer or sickle-cell anemia. What kinds of treatments have been proposed? How do they work? What are their limitations? Select the most interesting article and construct a brief report summarizing the principal outcomes.

5. Visit the library and research the latest developments in genetic engineering. How are scientists able to deliberately modify gene sequences in ways that are normally only possible in nature? How does gene replacement work?

6. Visit the library and research the effects of different environmental carcinogens (cancer-causing agents) on human beings. How do they exert their influence on the mechanisms that control cell growth and differentiation?

Challenge Questions

1. Why does protein synthesis take place on ribosomes? In what way does their structure support this function?

2. What is the advantage of synthesizing proteins from RNA—a modified copy of DNA—rather than from DNA itself?

3. You already know that some alterations in the genetic code can be harmful, even lethal. Can such alterations also be beneficial? Cite examples of alterations that might benefit an organism.

4. What is the role of mutation in altering the genetic code? Is this role significant in view of the other mechanisms for genetic alteration?

5. How does gene mutation differ from chromosomal variation?

6. How are scientists able to piece together the process of protein synthesis and the effects of genetic controls without direct observation?

7. Why is it easier to study gene control in prokaryotic cells than in eukaryotic cells?

8. Scientists have determined that eukaryotic cells use only a fraction of the DNA they contain. Why is this? Why is all this DNA passed down from one generation to the next?

9. Why do repeated applications of a single drug or pesticide cause resistance among some bacteria and insects?

LESSON 10

Microevolution

Assignments

For the most effective study of this lesson, we suggest that you complete the assignments in the sequence listed below:

Before Viewing the Video Program

- Read the Overview and Learning Objectives for this lesson. Use the Learning Objectives to guide your reading, viewing, and thinking.

- Read Chapter 16, "Microevolution," pages 238–259, and Chapter 17, "Speciation," pages 260–271, in the Starr textbook.

  Or read, in the Starr/Taggart textbook, Chapter 17, "Emergence of Evolutionary Thought," pages 270–279; Chapter 18, "Microevolution," pages 280–295; and Chapter 19, "Speciation," pages 296–309.

- Read the Viewing Notes in this lesson.

View the "Microevolution" Video Program

After Viewing the Video Program

- Briefly note your answers to the questions at the end of the Viewing Notes.

- Review all reading assignments for this lesson, especially the Chapter 16 and Chapter 17 Summaries on pages 258 and 270 in the Starr textbook (Chapter 17, Chapter 18, and Chapter 19 Summaries on pages 279, 294, and 308 respectively in the Starr/Taggart textbook) and the Viewing Notes in this lesson.

- Write brief answers to all the Review Questions at the end of Chapters 16 and 17 in the Starr textbook (Chapters 17, 18 and 19 in the Starr/Taggart textbook) to be certain you understand the text material.

- Complete the Review Activities in this guide to reinforce your understanding of important terms and concepts. Check your answers with the Answer Key and review when necessary.

- Take the Self-Test in this guide to measure your achievement of the Learning Objectives. Check your answers with the Answer Key and review when necessary.

- Complete the Using What You've Learned activities and any other activities and projects assigned by your instructor.

Overview

In the last lesson, you used your understanding of genetics and hereditary mechanisms to follow development of the components necessary to sustain individual cells and the larger organisms they constitute. In Lesson 10 you will rely on this same foundation to understand how populations of organisms change over time to exhibit different phenotypic characteristics. We refer, of course, to the process of evolution, the process which will be featured in this lesson and the next.

Lesson 10 will introduce you to the mechanisms of microevolution—mutation, gene flow, and natural selection among them—that change allele frequencies in populations over time. You'll learn how these mechanisms, when combined with reproductive isolation, give rise to new species. Later, in Lesson 11, you'll see how the processes of speciation and microevolution resulted in the large-scale patterns and trends that have characterized the history of life on this planet.

Your introduction to these concepts begins with the story of Charles Darwin and his five-year voyage on the *Beagle*. You'll learn of Darwin's discoveries on the Galápagos Islands and his studies of fossils in Argentina. You'll learn about other leading evolutionary theorists of the time and how they influenced Darwin's thinking. And you'll trace the development and ultimate publication of his ideas on natural selection in the now famous *Origin of Species* in 1859.

With this historical background as a prologue, you will begin to examine more closely the principle of natural selection and related concepts. You will learn that natural selection is not a conscious effort to perpetuate the species, but a passive mechanism for increasing the chance that individuals with traits best suited for a particular environment are able to pass on their alleles to the next generation. Variations of the natural selection process will be presented along with the conditions under which they occur.

In addition to natural selection, the lesson will present other important mechanisms for promoting the genetic divergence necessary for microevolution to occur. You'll learn about the role of mutation and genetic drift in changing the frequencies of alleles within a given population. You'll also learn how gene flow, the movement of alleles across population boundaries, contributes to genetic diversity. Through the use of historical and contemporary examples, these concepts will be brought to life as you continue the lesson.

Your understanding of these concepts will help you appreciate the expansion, diversification, and disappearance of familiar plants and animals over space and time. This is the subject of macroevolution, which will be the focus of Lesson 11.

Learning Objectives

When you have completed all assignments for this lesson, you should be able to:

1. Outline the major elements of the *theory of evolution by natural selection*, and identify the ideas and observations that influenced Darwin as he developed the theory.

2. Describe mutation and the other events that contribute to variation within a population.

3. Describe a population in terms of its *gene pool* and *allele (gene) frequency*, and outline the major conditions required to maintain *genetic equilibrium*.

4. Explain how *gene flow, genetic drift* due to population size, and *natural selection* can influence the rate and direction of changes in a population's allele frequencies.

5. Describe how patterns of environmental change and natural selection combine to shape a population's range of traits, and explain how these patterns influence the direction for evolution.

6. Define the term *species*, and describe the different isolation mechanisms that can promote *speciation*.

7. Describe how scientists study evolutionary changes in species.

Viewing Notes

The video program for Lesson 10 features Charles Darwin in a starring role as the father of evolutionary theory. With Alfred Wallace, Darwin threw the scientific community into turmoil with the publication of his *Origin of Species* in 1859. With the help of Dr. John Moore, Lesson 10 recalls the events and the observations that led Darwin to identify the primary mechanism of evolution, natural selection.

You will find the examples provided through this historical backdrop helpful in clarifying the concept of natural selection and how it works. You will learn that genetic variation is inherited in all populations and that conditions in the environment favor certain traits over others. It is this selection of adaptive traits rather than non-adaptive traits, over many generations, that drives the evolutionary process.

Later in the segment, Dr. Moore is joined by Dr. Ernst Mayr, who wrote a modern synthesis of evolutionary theory in the 1940s. Dr. Mayr helps bring Darwin's theories into focus, clarifying the essence of natural selection and microevolution in general.

Part Two of the program, "Evolution in Action," makes the transition from theory to practice as it follows the work of vector ecologists Richard Meyer and James Webb. Their study of vectors—disease-carrying organisms such as mosquitoes—helps to explain how populations adjust to catastrophic environmental changes. Dr. George Georghiou explains how the selective activation and deactivation of genes in the mosquito population can produce a pesticide-resistant strain over a period of several generations.

Through this example, you'll see how genetic equilibrium, once disrupted, provides the impetus for evolutionary change through mechanisms such as mutation, gene flow, and natural selection. The example also illustrates how human intervention can stimulate and accelerate natural evolutionary processes—for better or worse.

In Part Three, the process of natural selection is vividly demonstrated in the plants and animals indigenous to the small and geographically isolated island of Catalina. While naturalists Misty Gay and Allan Fone introduce you to the flora and fauna of the island, Dr. Ernst Mayr explains the mechanisms that promote the origin of new species in populations cut off from normal environmental pressures. You'll learn about concepts such as genetic drift and the founder effect and about patterns of reproductive behavior that isolate one population from another.

As you watch the video program, consider the following questions:

1. What lands and observations inspired Darwin to develop his theory of evolution?
2. How does the concept of natural selection explain variations in species descended from a common ancestor?
3. What was the initial response to Darwin's *Origin of Species* in 1859?

4. What is a vector, and what is the significance of vectors to the evolution of populations?
5. What other kinds of mechanisms can disrupt genetic equilibrium and evolutionary stability?
6. What techniques do scientists use to study microevolution?
7. What differentiates one species from another?
8. How do species originate?
9. What kinds of mechanisms tend to isolate populations enough to promote genetic divergence?

Review Activities

Matching

Match the terms listed below with the definitions that follow. Check your answers with the Answer Key and review any terms you missed.

I.

___ 1. allele frequency
___ 2. bottleneck
___ 3. catastrophism
___ 4. directional selection
___ 5. disruptive selection
___ 6. founder effect
___ 7. gene flow
___ 8. gene pool
___ 9. genetic drift
___ 10. genetic equilibrium
___ 11. inbreeding
___ 12. microevolution
___ 13. natural selection
___ 14. sexual selection
___ 15. stabilizing selection

a. all genotypes in a given population

b. a steady shift in allele frequencies toward one end of the range of phenotype variation

c. when the frequencies of alleles remain stable over generations

d. the relative number of each kind of allele carried by the totality of individuals constituting a population

e. differences in survival and reproduction among members of a population that tend to promote the persistence of some traits over others

f. non-random mating among closely related individuals who share many of the same alleles

g. a type of genetic drift in which a few individuals leave a population and establish a new one based on the limited alleles they carry

h. the process whereby changes in allele frequencies are brought about by genetic drift, gene flow, mutation, and natural selection

i. changes in allele frequencies due to the physical movement of alleles in and out of a population

j. persistence of the alleles responsible for the most common phenotypes

k. the change in allele frequencies that occurs over generations due only to chance; it is especially important in small populations

l. a type of natural selection that favors a trait giving an individual a reproductive advantage

Lesson 10 / Microevolution   121

m. a shift in allele frequencies due to extremes of phenotypic variation and away from intermediate forms

n. a severe reduction in population size brought about by a catastrophic event or intense pressure for natural selection

o. theory of evolution that a global event caused the extinction of many species, allowing the survivors to repopulate the world

II.

___ 1. adaptive radiation
___ 2. adaptive zone
___ 3. allopatric speciation
___ 4. anagenesis
___ 5. background extinction
___ 6. cladogenesis
___ 7. hybrid zone
___ 8. mass extinction
___ 9. parapatric speciation
___ 10. speciation
___ 11. sympatric speciation

a. speciation that occurs when accumulated changes in allele frequencies occur in a single, unbranched line of descent

b. speciation that occurs when two populations share a common border that allows gene flow

c. a sudden rise in extinction rates above normal background extinction, usually due to a catastrophic event

d. expected rate of species disappearance over time

e. a way of life that is acquired by a lineage with physical, ecological, and evolutionary access to it

f. the burst of speciation that occurs when lineages branch out into new environments

g. speciation that occurs when populations become isolated and genetically branch in different directions

h. speciation that occurs within an existing population in the absence of physical barriers

i. an area where adjoining populations meet, interbreed, and produce hybrid offspring

j. speciation that occurs when populations of the same species become geographically isolated

k. the process by which diverging populations accumulate enough differences in allele frequencies to prevent interbreeding

Completion

Fill in each blank with the most appropriate term from the list for that paragraph. A term may be used once, more than once, or not at all. If a question requires two or more answers in succession, they may be in any order (unless the question indicates otherwise). Check your answers with the Answer Key and review when necessary.

1. Studies involving _____, _____, and _____ all provide clues to understanding evolution. Evolutionary tree diagrams of some lineages show small morphological changes over long spans of time, supporting a _____. Other lineages show more abrupt morphological changes in a relatively brief amount of time, supporting the _____. Species that are able to exploit an _____ show accelerated evolutionary activity (_____). Species that are unsuccessful in adapting to local change may become _____. When a catastrophic event causes species to disappear faster than _____, _____ occurs.

adaptive radiation	extinct
adaptive zone	fossils
background extinction	gradual model of speciation
biogeography	mass extinction
comparative morphology	punctuation model of speciation

2. _____ occurs when _____ frequencies in a population change over time. These changes can result from the creation of new alleles through _____, from the retention of favorable alleles through _____, from the physical movement of alleles into and out of the population (_____), and from random changes (_____). The effect of genetic drift is particularly strong with _____ and _____ populations. Populations which are not evolving through these mechanisms are at _____.

allele	genetic drift
anagenesis	genetic equilibrium
bottlenecked	inbred
emigration	microevolution
extinct	mutation
gene flow	natural selection
gene pool	

3. Sexually reproducing populations that can successfully _____ constitute a _____. _____ have a common genetic history and evolve independently of other species. When a subpopulation builds up _____ differences through _____, _____, _____, or _____, _____ may produce a new species. Some of the mechanisms that isolate populations prior to _____ include differences in reproductive timing and behavior, incompatibilities in reproductive structures, sterility, early death, and geographic separations.

    allelic   mutation

    gene flow   natural selection

    genetic divergence   speciation

    genetic drift   species

    interbreed

Self-Test

Select the one best answer for each question. Check your answers with the Answer Key and review when necessary.

1. Comparisons of the fossil remains of glyptodonts with modern-day armadillos led Darwin to conclude that

    a. offspring inherit physical changes developed by parents over their lifetime in response to internal needs and environmental pressures.

    b. geologic forces that reshaped Earth's surface millions of years ago are still operating today.

    c. variations in inherited traits give some organisms an adaptive advantage in certain environments.

    d. a population is naturally evolving when the number of helpful traits in the gene pool equals the number of harmful traits.

2. Which of the following mechanisms is the only one that creates *new* alleles in a population?

    a. translocation

    b. mutation

    c. fertilization

    d. independent assortment

3. Genetic equilibrium is achieved when

    a. the allele frequencies at a given gene locus are stable over successive generations.

    b. alleles that produce useful traits outnumber alleles that produce harmful traits.

    c. there is an equivalent exchange of alleles between one population and another.

    d. there is no intermixing of alleles through fertilization or independent assortment.

4. Which of the following conditions must be met for a population to achieve genetic equilibrium?

    a. There can be only one of each kind of allele in the gene pool.

    b. The population must be very small in number.

    c. New alleles produced through mutations must not affect reproduction.

    d. The population must be isolated from other populations of the same species.

5. Which of the following probably accounts for most of the morphological changes that have occurred throughout the history of life?
   a. mutation
   b. genetic drift
   c. gene flow
   d. natural selection

6. Genetic drift has a significant effect on allele frequencies in
   a. very large populations.
   b. very small populations.
   c. populations that are breeding with other populations.
   d. populations that are genetically homogeneous.

7. Which of the following conditions would promote directional selection?
   a. a steady, significant change in environmental conditions
   b. repetitive, seasonal cycles of environmental change
   c. environmental conditions that remain unchanged for a long period
   d. a sudden, cataclysmic change in environmental conditions

8. The most common mechanism of speciation is
   a. allopatric speciation.
   b. sympatric speciation.
   c. parapatric speciation.
   d. omnipatric speciation.

9. The study of resistance to chemicals such as antibiotics and pesticides provides scientists with clues to
   a. the rate of mutation among different populations.
   b. the role of genetic drift in microevolution.
   c. how directional selection works.
   d. why members of some populations react more quickly to environmental change.

Using What You've Learned

Based on your own interests or your instructor's requirements, complete one or more of the following activities.

1. Identify a particular species of animal or plant and research its evolution to the present day. Use maps, evolutionary tree diagrams, and other visual aids to show how environmental changes and the forces of microevolution (i.e., speciation, adaptive radiation, natural selection, gene flow, etc.) contributed to its current status.

2. As a variation of item 1, research a particular disease that is associated with a particular group (e.g., Tay-Sachs disease among Ashkenazic Jews, sickle-cell anemia in African Americans, TB among Asians). Identify the principal mechanisms that led to the spread of the disease within the population and prevent it from spreading to other populations.

3. Assume for a moment that the neighborhood you live in was permanently transported to a remote island. Speculate on what would happen to allele frequencies and their phenotypic expression over several generations of interbreeding. What kinds of traits would be promoted? Which traits would be lost? What role might environment play in the evolution of your population?

4. Based on your understanding of the mechanisms of microevolution, speculate on how human beings might evolve over time given various environmental scenarios. For example, how might we evolve as a result of a natural or human-caused catastrophe? What about more gradual environmental or social changes? Develop at least three different scenarios and explain the rationale for your predictions.

5. Visit your library and research the methods scientists use to perpetuate endangered populations. What special techniques do they use? What problems do they encounter? What is the success rate? If you wish, select a specific example of an endangered species and find out more about why the species is endangered and what is being done about it.

Challenge Questions

1. Which evolutionary model is best supported by the fossil record—the gradual or the punctuation model? What kind of evidence supports each model?

2. How important is the creation of new alleles to the evolutionary process? What would happen if there were no genetic mutations?

3. How might inbreeding affect the evolution of species over successive generations?

4. Is it better for populations to maintain genetic diversity or should they strive to become genetically homogeneous? Explain.

5. Why do island populations such as those of Hawaii or the Galápagos Islands often display a wide variety of phenotypes within a given species?

6. What accounts for the distinctive patterns and colors between males and females of the same species?

7. Why is it that some individuals can roll their tongues and others can't? Why do alleles for both traits persist in the gene pool?

8. Why is it that in a given population, many more offspring are produced than there are resources to support them?

LESSON 11

Macroevolution

Assignments

For the most effective study of this lesson, we suggest that you complete the assignments in the sequence listed below:

Before Viewing the Video Program

- Read the Overview and Learning Objectives for this lesson. Use the Learning Objectives to guide your reading, viewing, and thinking.

- Read Chapter 18, "The Macroevolutionary Puzzle," pages 272–288, and Chapter 19, "The Origin and Evolution of Life," pages 290–309, in the Starr textbook.

  Or read, in the Starr/Taggart textbook, Chapter 20, "The Macroevolutionary Puzzle," pages 310–330, and Chapter 21, "The Origin and Evolution of Life," pages 332–351.

- Read the Viewing Notes in this lesson.

View the "Macroevolution" Video Program

After Viewing the Video Program

- Briefly note your answers to the questions at the end of the Viewing Notes.

- Review all reading assignments for this lesson, especially the Chapter 18 and Chapter 19 Summaries on pages 287 and 308 in the Starr textbook (Chapter 20 and Chapter 21 Summaries on pages 329 and 350 respectively in the Starr/Taggart textbook) and the Viewing Notes in this lesson.

- Write brief answers to all the Review Questions at the end of Chapters 18 and 19 in the Starr textbook (Chapters 20 and 21 in the Starr/Taggart textbook) to be certain you understand the text material.

- Complete the Review Activities in this guide to reinforce your understanding of important terms and concepts. Check your answers with the Answer Key and review when necessary.

- Take the Self-Test in this guide to measure your achievement of the Learning Objectives. Check your answers with the Answer Key and review when necessary.

- Complete the Using What You've Learned activities and any other activities and projects assigned by your instructor.

Overview

In Lesson 10, you learned how the genetic mechanisms and hereditary principles discussed in earlier lessons interact with environmental conditions to determine the survival of entire populations. The dynamics of natural selection, outlined almost 140 years ago by Charles Darwin, provide an important foundation for understanding how life evolved on planet Earth.

Lesson 11 takes up the challenge by describing how life might have originated and diversified into the plethora of species we know today. It paints a picture of primordial Earth, an uninviting place that held the essential ingredients and conditions for creating carbon-based life. It describes a billion-year process during which organic molecules combined to form amino acids, then proteins, then primitive prokaryotic cells—the forerunners of today's bacteria. The lesson also explains how oxygen, an essential requirement for so many life forms today, was instrumental in stimulating the evolution of eukaryotic cells, which form the basis for multicellular organisms today.

You'll learn how life has branched out over the last four billion years to fill the five major kingdoms we use to classify life on Earth: prokaryotes, eukaryotes, fungi, plants, and animals. In the process, you'll explore the ups and downs of prehistoric organisms as they evolved through the major divisions of geologic time—from the Protozoic era to the Cenezoic era—and the kinds of catastrophic changes that led to the extinction and expansion of populations on a global scale.

While the story of evolution is a fascinating one, it is equally intriguing to learn about the methods scientists use to solve the macroevolutionary puzzle. Lesson 11 will give you some insight into the methods paleontologists, biochemists, and other researchers use to reconstruct the events and processes that determined the paths different organisms have traveled over time. You'll learn, for example, how fossils are used to trace species to a common ancestor and allow scientists to speculate on global conditions at different points in history. You'll see how comparative studies of cellular proteins and embryonic development yield clues to the common origins of different species, including human beings.

Learning Objectives

When you have completed all assignments for this lesson, you should be able to:

1. List what biologists generally accept as evidence to support the theory of evolution, and explain how *paleontology, comparative morphology, comparative biochemistry*, and other techniques and observations are used to reconstruct the past.

2. Contrast the primordial Earth atmosphere with that of today, and describe the modern experiments that suggest how life may have originated on Earth.

3. Discuss the reasons why, of all the planets in our solar system, only Earth is uniquely adapted to support life as we know it.

4. Describe the general movement of tectonic plates in the *Paleozoic, Mesozoic,* and *Cenozoic eras*, and relate this movement to changes in the fossil record.

5. Discuss the factors that encourage increased rates of speciation and the formation of larger taxonomic groups, and identify the factors that bring about extinction and replacement of species.

6. Identify some of the principal organisms associated with each of the five geologic eras.

Viewing Notes

"Macroevolution" will take you on a fantastic journey through time to witness the birth of our planet and life as we know it. The journey begins with the formation of the Earth from a molten mass of gas and rock, followed by a billion years of hot gases and surface turbulence in the form of volcanic eruptions. As the Earth cooled and water condensed into rainfall to form the oceans of the world, conditions became ripe for the first forms of life.

Part One of the video vividly traces these and other crucial developments in the evolution of life. You'll see how modern-day paleontologists such as Dr. Springer and Dr. Scott have pieced the macroevolutionary puzzle together using the fossil record and other techniques. You'll see how scientists have been able to use the fossil record to make assumptions not only about what different species looked like, but how closely related they are and how they have changed over time to adapt to changing conditions on the planet.

Your journey will take you to one of the richest repositories of fossils from the Cenozoic era—the Rancho de la Brea Tar Pits. Dr. Blaire Van Valkenburgh will show you how he and other paleontologists reconstruct the appearance, the behavior, even the environment in which prehistoric beasts, such as the saber-toothed tiger, lived.

In Part Two of the video program, your journey will take you to even deeper levels of the fossil record to learn how climatic changes stimulated population migration and diversification in the Paleozoic and Mesozoic eras. Paleoecologist Dr. David Bottjer views these sweeping changes as natural evolutionary processes, processes which tend to occur in fits and starts to match cataclysmic changes on Earth. You'll find out why mass extinctions occurred and what it meant to the evolutionary success of the survivors.

In Part Three, you will go back even further to a time when life on Earth consisted of primitive one-celled organisms, and there were no bones or other body structures to fossilize. Without fossils to study, Dr. Stanley Miller and Dr. Russell Doolittle describe the methods scientists have used to reconstruct those events leading to the formation of amino acids and the first unicellular organism.

They describe the milestone experiment in which graduate student Stanley Miller recreated the atmospheric conditions of primitive Earth in a laboratory and succeeded in forming amino acids—one of the building blocks of life. They also describe the important role oxygen—or rather the lack of oxygen—played in the evolution of primitive cells from these building blocks. Dr. Doolittle explains how biochemical comparisons of protein sequences are used to establish the family trees of even the earliest single-celled organisms and how we may have more in common with bacteria than we think.

As you watch the video program, consider the following questions:

1. How do fossils help scientists piece together the "family tree" of a given plant or animal?
2. How do fossils provide evidence for the theory of continental drift?

3. What are the two major forces that drive the evolutionary process?

4. Why do species become extinct? What does extinction of one species mean to the surviving species in the same environment?

5. Why are there "gaps" in the fossil record, particularly at the end of the Paleozoic and Mesozoic eras?

6. What stimulated the evolution of mammals at the end of the Mesozoic era?

7. What was the significance of Miller's experiment in solving the macroevolutionary puzzle?

8. How did the absence of oxygen stimulate the evolution of life on Earth?

9. How does comparative biochemistry help establish the lineage of even the most primordial organisms?

10. In what ways are we similar to primitive single-celled organisms such as bacteria? What does this suggest in terms of our evolution?

Review Activities

Matching

Match the terms listed below with the definitions that follow. Check your answers with the Answer Key and review any terms you missed.

I.

___ 1. analogous structures
___ 2. biological clock
___ 3. comparative morphology
___ 4. DNA-DNA hybridization
___ 5. fossilization
___ 6. homologous structures
___ 7. macroevolution
___ 8. molecular clock
___ 9. neutral mutation
___ 10. phylogeny
___ 11. stratification
___ 12. taxa

a. a gene mutation that has neither a helpful nor a harmful effect on the survival of the organism

b. groupings of species for the purposes of classification

c. body structures that developed in separate lineages, but resemble one another because they were put to comparable uses in similar environments

d. layering of sedimentary deposits over time

e. gradual process whereby the remains of an organism become buried and solidified due to fusion with surrounding inorganic compounds

f. body structures that evolved from a common ancestor, but developed in different ways in different lineages

g. base pairing of nucleotide sequences from different species for the purpose of evolutionary studies or genetic engineering

h. evolutionary relationships among species, including their ancestral roots and the branches of their descendants

i. internal mechanism that helps an organism adjust to daily and seasonal activities based on environmental cues

j. a method of estimating the time of origin of species or lineages based on the presumed role of accumulation of neutral mutations

k. large-scale patterns, trends, and rates of change among species

l. comparisons of the anatomical structures of major lineages

II.

___ 1. archaebacterium   ___ 6. Mesozoic
___ 2. Archean   ___ 7. Paleozoic
___ 3. Cenozoic   ___ 8. Proterozoic
___ 4. endosymbiosis   ___ 9. protistan
___ 5. eubacterium   ___ 10. stromatolite

a. mutually beneficial interdependency between two species

b. geologic era characterized by expansion of all kingdoms in the seas and initial invasion of land environments

c. geologic era characterized by great geologic and climatic changes, and the rise of mammals

d. layered underwater structures formed from sediments and remains of photosynthetic organisms

e. one of the three prokaryotic lineages that arose early in evolution, now representing halophiles, methanogens, and thermophiles

f. first single-celled eukaryote

g. geologic era characterized by increasing global diversity and rise of the dinosaurs

h. geologic era characterized by the divergence of prokaryotic organisms

i. geologic era characterized by the evolution of eukaryotic cells

j. one of the three original prokaryotic lineages, includes all bacteria except halophiles, methanogens, and thermophiles

Completion

Fill in each blank with the most appropriate term from the list for that paragraph. A term may be used once, more than once, or not at all. If a question requires two or more answers in succession, they may be in any order (unless the question indicates otherwise). Check your answers with the Answer Key and review when necessary.

1. _____ evidence for the evolution of life on Earth comes from many sources. Chemical analysis of the _____ and _____ suggests that the precursors of _____ were available. Experiments have shown that conditions on primordial Earth could have produced the _____, _____, and _____, which are the essential ingredients of living organisms. Based on chemical principles and _____, it appears that the metabolic pathways of modern-day cells originated through chemical competition for _____. Self-replicating systems of _____ and _____ have been artificially reproduced in the laboratory.

amino acids	indirect
atmosphere	lipid protein membranes
organic molecules	oceans
computer simulations	organelles
direct	other planets
DNA	RNA
Earth	sugars
enzyme(s)	sun

2. Life began in the Archean era when the first _____ bacteria appeared. From that beginning arose three major lineages of single-celled organisms: _____, _____, and the ancestors of _____ cells. While some of these organisms used a _____ pathway of photosynthesis, others used a _____ pathway which produced _____ as a waste product. The accumulation of free _____ eventually prevented the spontaneous formation of organic molecules and stimulated the evolution of _____ respiration. The endosymbiosis between early _____ bacteria and the _____ ancestors of cells probably led to the forerunners of today's _____ and _____. The _____-rich atmosphere also provided enough protection from _____ to allow some forms of sea life to move onto land.

aerobic	high temperature
anaerobic	mitochondria
archaebacteria	nitrogen
carbon dioxide	non-cyclic
chloroplasts	oxygen
cyclic	prokaryotic
eubacteria	ultraviolet radiation
eukaryotic	unstable weather conditions

3. _____ and _____ reveal the patterns of changes that have characterized _____ on this planet. Interpretation of the _____ record, _____ comparisons of major lineages, and observations of _____ development all provide evidence of common ancestral roots. Protein and _____ comparisons give us an indication of when major lineages began. These lineages are named according to a _____ classification scheme developed by _____. This scheme was eventually expanded into the _____ scheme we use today. The five major kingdoms in this system are _____ (prokaryotes), _____ (eukaryotes), _____ (heterotrophs), _____ (plants), and _____ (animals).

anatomical	fossil
Animalia	Fungi
binomial	Linnaeus
comparative biochemistry	macroevolution
comparative morphology	Monera
da Vinci	nucleic acid
Darwin	phylogenetic
embryonic	Plantae
endosymbiosis	Protista

Lesson 11 / Macroevolution 137

Self-Test

Select the one best answer for each question. Check your answers with the Answer Key and review when necessary.

1. Which of the following provides evidence for the theory of evolution?

    a. mathematical projections of spontaneous chemical reactions

    b. theories of plate tectonics and continental drift

    c. the creation of new species through accelerated genetic mutation

    d. gradual morphological changes in the fossil record

2. Comparative biochemistry tells us that

    a. only 25% of genetic material is common to all species.

    b. organisms rapidly diversified after the first prokaryotic cell appeared.

    c. at some point in the past, humans and bacteria had a similar ancestor.

    d. species in different locations evolved similar genetic patterns to adapt to similar environments.

3. Before Earth's crust cooled and solidified, the atmosphere consisted primarily of

    a. methane and ammonia gas.

    b. carbon dioxide and water vapor.

    c. hydrogen, nitrogen, carbon monoxide, and carbon dioxide.

    d. sulfurous compounds from volcanic eruptions.

4. When Stanley Miller introduced an electrical discharge into a gas-filled reaction chamber he demonstrated how

    a. the oceans became the spawning ground for primitive cells.

    b. amino acids could spontaneously form under Earthlike conditions.

    c. the atmosphere of the Earth changed to its present composition.

    d. life could evolve from simple organic compounds.

5. Of all the planets in the solar system, Earth was uniquely adapted to support life as we know it because

    a. its surface gravity was neither too low nor too high.

    b. it received just the right amount of sunlight.

    c. its atmosphere contained methane and ammonia.

    d. its temperature allowed water to remain in a liquid state.

6. The discovery of mesosaur fossils in Brazil and the West Coast of Africa is evidence of

    a. plate tectonics.

    b. morphological convergence.

    c. evolutionary migration.

    d. adaptive radiation.

7. The collision of land masses to form Pangea, the supercontinent that extended from pole to pole, occurred

    a. at the beginning of the Proterozoic era.

    b. at the end of the Paleozoic era.

    c. during the Mesozoic era.

    d. at the end of the Cenozoic era.

8. Mass extinctions of prehistoric life appear to be the result of

    a. cataclysmic geologic or climatic changes.

    b. dramatic changes in genetic mutation rates.

    c. loss of adaptability due to an extremely limited gene pool.

    d. biological plague.

9. In which of the following eras would you expect to find a saber-toothed tiger?

    a. Paleozoic

    b. Cenozoic

    c. Mesozoic

    d. Proterozoic

10. What was the dominant form of life during the Archean era?

    a. trilobites

    b. prokaryotes

    c. flowering plants

    d. small land mammals

Using What You've Learned

Based on your own interests or your instructor's requirements, complete one or more of the following activities.

1. Using charts and other illustrations, trace the evolution of one species or group of animals or plants over time.

2. Prepare a timeline describing the geological and biological history of your local region. Illustrate the kinds of plants and animals that were prevalent during specific eras, periods, and epochs.

3. Visit your local or regional natural history museum. What kinds of fossils have been found in your area? What do they reveal about the natural history of the region?

4. Visit the library and scan recent science publications for articles on species extinction. What is the rate of extinction today? How is it different from the background extinction rate one would expect? What areas are experiencing the greatest rate of extinction and why?

5. Consider how life might evolve on other planets such as Venus or Mars. Select a planet in our solar system and research its surface conditions, atmosphere, temperature, gravity, etc. Speculate on whether life might develop under such conditions and what forms it might take.

6. Read Chapter 24 in the Starr textbook (Chapter 28 in the Starr/Taggart textbook) to get an idea of how the phylogenetic classification system is used to classify members of the kingdom Animalia. If possible, visit a zoo, noting how the various phyla, classes, orders, families, genera, and species within the animal kingdom diversified over millions of years of evolution.

7. Chapter 24 in the Starr textbook (Chapter 28 in the Starr/Taggart textbook) also discusses the evolution of humans or *Homo sapiens*. Present opposing viewpoints on the origins of humans, representing current creationist and scientific arguments.

Challenge Questions

1. Fossils are critical to solving the macroevolutionary puzzle. How is the age of fossils determined?

2. What kinds of organisms are typically found in the fossil record? What kinds of organisms are underrepresented, and what other methods do scientists use to learn about their evolution?

3. What is the difference between microevolution and macroevolution? What are some examples of each?

4. What kind of evidence suggests that evolution occurs in bursts of great diversification rather than through a steady, gradual process of change? What triggers these evolutionary spurts?

5. Most biologists accept Darwin's theory of natural selection or "survival of the fittest" as an explanation for why species thrive or die out. Does this theory explain the evolution of complex organic compounds into primitive, single-cellular organisms, and later into multicellular organisms? Why or why not?

6. Where did the oxygen in our atmosphere come from, and how did it affect the evolution of life on our planet?

7. Why do some species appear very much as they did millions of years ago, while others have radically changed?

8. Does the story of macroevolution on this planet lend support to the possibility of life evolving on other planets? Why or why not?

9. When we think of animals, most of us think of chordates (organisms with a spinal cord). Yet chordates represent only 5% of the species in the animal kingdom. What other kinds of organisms are classified as "animals"?

10. What kinds of methods have scientists used to piece together the evolution of *Homo sapiens*?

LESSON
12

Viruses, Bacteria, and Protistans

Assignments

For the most effective study of this lesson, we suggest that you complete the assignments in the sequence listed below:

Before Viewing the Video Program

- Read the Overview and Learning Objectives for this lesson. Use the Learning Objectives to guide your reading, viewing, and thinking.

- Read Chapter 20, "Bacteria, Viruses, and Protistans," pages 310–331, in the Starr textbook. Also review Chapter 18, "The Macroevolutionary Puzzle," pages 282–283, and Chapter 19, "The Origin and Evolution of Life," pages 298–299, in the Starr textbook.

    Or read, in the Starr/Taggart textbook, Chapter 22, "Bacteria and Viruses," pages 352–369, and Chapter 23, "Protistans," pages 370–387. Also review Chapter 20, "The Macroevolutionary Puzzle," pages 322–329, and Chapter 21, "The Origin and Evolution of Life," pages 338–339.

- Read the Viewing Notes in this lesson.

View the "Viruses, Bacteria, and Protistans" Video Program
After Viewing the Video Program

- Briefly note your answers to the questions at the end of the Viewing Notes.

- Review all reading assignments for this lesson, especially the Chapter 20 Summary on page 330 in the Starr textbook (Chapter 22 and Chapter 23 Summaries on pages 368 and 386 respectively in the Starr/Taggart textbook) and the Viewing Notes in this lesson.

- Write brief answers to all the Review Questions at the end of Chapter 20 in the Starr textbook (Chapters 22 and 23 in the Starr/Taggart textbook) to be certain you understand the text material.

- Complete the Review Activities in this guide to reinforce your understanding of important terms and concepts. Check your answers with the Answer Key and review when necessary.

- Take the Self-Test in this guide to measure your achievement of the Learning Objectives. Check your answers with the Answer Key and review when necessary.

- Complete the Using What You've Learned activities and any other activities and projects assigned by your instructor.

Overview

Lesson 12 introduces one of the true marvels of biological science: a classification system that attempts to provide a phylogenetic link between every organism and every other organism on the planet. As you'll discover, the original scheme developed by Linnaeus in the eighteenth century has undergone many revisions, ultimately yielding the five-kingdom classification system we use today.

Having established the dividing lines among the kingdoms Bacteria, Protista, Fungi, Plantae, and Animalia, Lesson 12 will introduce you to members of the first two kingdoms, Bacteria and Protista. These include the simplest of organisms, the single-celled prokaryotes and their eukaryotic descendants. You'll learn about the characteristics that distinguish each of these ancient groups and the morphological variations that have arisen over eons of evolutionary development. While the lesson emphasizes the pathogenic effects of many species of bacteria and protistans, you will also learn about species that have a beneficial effect on humankind.

As a bonus, Lesson 12 will also introduce you to a special group of infectious agents that don't even qualify as living organisms—the viruses. Although technically outside of the biological classification scheme, viruses—like bacteria and protistans—have a big impact on human health and well-being. Witness the effects of influenza, herpes, and AIDS. As you will soon discover, it is the very simplicity of these agents that enables them to avoid detection and eradication through traditional medical regimens.

Learning Objectives

When you have completed all assignments for this lesson, you should be able to:

1. State the purpose of a *classification system,* and list—in sequence—the major units or groupings of the system we use today. Understand the difficulty of classifying organisms into any artificially derived scheme.

2. Name the five kingdoms into which living organisms are presently classified, and explain why viruses are not included in the scheme.

3. Identify the general characteristics of a virus, describe the basic steps of viral replication, and list examples of human illnesses caused by viral infection.

4. Describe the metabolic processes, structural features and reproductive patterns of bacteria, and distinguish among the main groups.

5. Describe the characteristics unique to protistans, and distinguish between the major phyla.

6. Summarize the theory of *endosymbiosis,* and use a phylogenetic tree to explain the possible evolutionary relationships among the five kingdoms.

Viewing Notes

Your video survey of bacteria, viruses, and protistans begins with a history of biological classification systems. You'll learn how humankind, from as far back as the ancient Greeks, has tried to impose some kind of logical organization on the vast diversity of life on this planet. From Dr. James Lake and Dr. Russell Doolittle, you'll learn about the binomial system biologists use today to categorize different species and how this system provides clues to the phylogenetic origin of different kinds of plants and animals. More important, you'll begin to understand the nature and origin of the similarities and differences between the various organisms and how they came about.

Part Two of the video program explains the potential health threats posed by a particular group of organisms—bacteria. Dr. Hildy Meyers, Medical Director of Communicable Disease Control and Epidemiology for Orange County, California, describes the various bacterial threats to the community through food and other sources. With this backdrop, you'll trace the evolution of the kingdom Monera, which includes prokaryotes such as the archaebacteria and eubacteria, and learn about the fundamental characteristics of these organisms. You'll actually observe the structure and reproduction of prokaryotic cells, courtesy of computer animation. Returning to the public health theme, you'll see how epidemiologists study bacteria in an effort to develop effective antibiotics.

As Part Two continues, you'll learn of a group of infectious agents that are completely resistant to antibiotic treatment—viruses. You'll find out how these simple collections of nucleic acids inside protein shells defy classification as living creatures, yet are able to commandeer the genetic machinery of a living cell to reproduce themselves. Computer animation and crystalline visualization will show you how viruses are identified. You'll also learn about the public health risks posed by viruses and what science is doing to defend us against them.

Part Three introduces the protistans, which are also surveyed in Lesson 12. With a kingdom all to themselves, protistans include the plankton and dinoflagellates of oceans, the protozoa often found in freshwater ponds, and the multicellular colonies of algae and seaweed. Dr. Peter Franks of the Scripps Institute of Oceanography is your guide for this segment of the program. He describes the ancestry of the eukaryotic protistans, their common characteristics, and the concept of endosymbiosis, which may explain how they evolved from guest-host relationships with more primitive prokaryotic cells.

As you watch the video program, consider the following questions:

1. What is the basis for dividing living organisms into the five major kingdoms used for biological classification today?
2. What kinds of recent classification techniques are changing the way species are categorized and understood by scientists?
3. Why is it important to understand the lineage and history of an organism or group of organisms?

4. What kinds of bacteria are commonly found in foods? Are any of these bacteria potentially harmful?

5. How do archaebacteria and eubacteria differ? To which group do most contemporary bacteria belong?

6. Why aren't viruses considered living organisms?

7. How do viruses replicate themselves?

8. What health risks are posed by viruses, and how is science helping to provide a defense against these invaders?

9. What are protistans, and what distinguishes them as a separate kingdom?

10. How did eukaryotes evolve from prokaryotes?

Review Activities

Matching

Match the terms listed below with the definitions that follow. Check your answers with the Answer Key and review any terms you missed.

I.

___ 1. archaebacterium   ___ 7. moneran

___ 2. bacteriophage   ___ 8. phytoplankton

___ 3. chytrid   ___ 9. protozoan

___ 4. eubacterium   ___ 10. sporozoan

___ 5. halophile   ___ 11. thermophile (extreme)

___ 6. methanogen   ___ 12. virus

a. a virus that infects bacterial cells

b. a single-celled eukaryote that resembles the heterotrophs that gave rise to animals

c. a single-celled parasite that produces sporelike infectious agents

d. a community of floating or swimming photosynthetic autotrophs

e. an archaebacterium that lives in very salty habitats

f. an archaebacterium that lives in or around hot springs, acidic soils, and hydrothermal vents

g. a single-celled prokaryote; a bacterium

h. a type of protist that lives in fresh water and marine habitats

i. any of the single-celled halophiles, extreme thermophiles, and methanogens

j. any bacterium other than a halophile, extreme thermophile, or methanogen

k. an archaebacterium that lives in oxygen-free habitats and produces methane gas as a by-product

l. a noncellular infectious agent that uses a host cell's metabolic machinery to sustain and replicate itself

Now match the examples listed below with the appropriate taxonomic level. Check your answers with the Answer Key and review any classification you missed.

II.

___ 1. Kingdom ___ 5. Family

___ 2. Phylum ___ 6. Genus

___ 3. Class ___ 7. Species

___ 4. Order

a. Example: Chordata

b. Example: *Homo*

c. Example: Primates

d. Example: Animalia

e. Example: *sapiens*

f. Example: Mammalia

g. Example: Hominidae

Completion

Fill in each blank with the most appropriate term from the list for that paragraph. A term may be used once, more than once, or not at all. If a question requires two or more answers in succession, they may be in any order (unless the question indicates otherwise). Check your answers with the Answer Key and review when necessary.

1. The _____ classification scheme developed by _____ uses a generic descriptor (_____) with a specific name to identify each unique kind of organism. The scheme is based on _____ and consists of _____ major _____. One of these is the kingdom Monera, which includes the _____. Another is the kingdom Protista, which includes eukaryotes such as _____, _____, and _____.

   kingdoms   three
   morphology   five
   phyla   seven
   phylogeny   Linnaeus
   water molds   Darwin
   genus   viruses
   polynomial   bacteria
   monomial   order
   binomial   sporozoans
   protozoa

2. Bacteria are _____ cells that come in various shapes. Some are spherical (_____), some are rod-like (bacilli), and some are spiral. They contain _____ and _____, but have none of the organelles common to eukaryotic cells. They move by using a _____, which rotates like a propeller, and reproduce by _____. While some bacteria use sunlight and carbon dioxide to produce sugar through photosynthesis, other bacteria use carbon dioxide alone. _____ bacteria get their energy by feeding off living hosts or organic wastes.

   prokaryotic   DNA
   eukaryotic   mitosis
   binary fission   meiosis
   flagellum   cocci
   mitochondria   pseudopod
   chemoautotrophs   ribosomes
   chemoheterotrophic   a nucleus
   photoheterotrophs

Lesson 12 / Viruses, Bacteria, and Protistans   149

3. Viruses are _____, _____ infectious agents with two distinctive traits. First, they have a core of _____ or _____ covered with a _____ coat. The coat is often enclosed within a _____ envelope that has _____ or _____ spikes or projections. Second, they inject their _____ into a _____ cell where it directs the cell to produce viral _____ and _____. These components are assembled into new viral particles that _____ the infected cell.

protein(s)	single-celled
carbohydrate	noncellular
glucose	consume
lipid	are released from
host	take over
living	DNA
nonliving	RNA
dormant	

150 Lesson 12 / Viruses, Bacteria, and Protistans

Self-Test

Select the one best answer for each question. Check your answers with the Answer Key and review when necessary.

1. Classification systems like the current binomial system are useful for
   a. determining which organisms are likely to survive.
   b. describing the distinguishing features of each organism.
   c. showing the evolutionary relationships among species.
   d. identifying which species have been studied and which have not.

2. In the current classification system, the next taxonomic level following kingdom is
   a. genus.
   b. phylum.
   c. order.
   d. family.

3. Which of the following is **NOT** one of five major kingdoms into which living organisms are classified?
   a. Plantae
   b. Monera
   c. Insertivera
   d. Protistan

4. Viruses are not included in the current scheme used to classify living organisms because
   a. they are not considered living organisms.
   b. they do not fit into any established kingdom.
   c. some of them are like eukaryotes, others like prokaryotes.
   d. they were discovered only recently.

5. Viruses are able to replicate themselves by
   a. binary fission.
   b. assembling free-floating nucleic acids into viral packages.
   c. converting host cells into viral particles.
   d. directing the host cell to copy viral DNA and proteins.

6. Which of the following diseases is believed to be caused by a virus?
    a. streptococcus
    b. melanoma
    c. influenza
    d. salmonella

7. Which of the following features is characteristic of most bacteria?
    a. They have a membrane-bound nucleus and primitive organelles.
    b. They have a cell wall which consists of peptidoglycan.
    c. They are parasites, feeding off the glucose of a living host.
    d. They reproduce by meiosis.

8. Bacteria such as *E. coli* and *Clostridium* are classified as
    a. photoautotrophic eubacteria.
    b. photoheterotrophic eubacteria.
    c. chemoautotrophic eubacteria.
    d. chemoheterotrophic eubacteria.

9. One common trait among protistans is that they all
    a. have a nucleus, Golgi bodies, and mitochondria.
    b. are single-celled.
    c. are heterotrophs.
    d. use pseudopods, cilia, or flagella for movement.

10. The theory of endosymbiosis explains how
    a. some bacteria develop parasitic relationships with other organisms.
    b. prokaryotes lost the organelles currently found in eukaryotes.
    c. protistans diverged from monerans on the evolutionary tree.
    d. plants and animals use the same mechanisms for obtaining and using energy.

Using What You've Learned

Based on your own interests or your instructor's requirements, complete one or more of the following activities.

1. Create a three-column chart showing the kinds of human diseases caused by bacteria, viruses, and protozoa. Which diseases are most serious? Which diseases are most difficult to treat? What are the recommended treatments in each case?

2. Investigate the work of the World Health Organization and the Centers for Disease Control in the study and treatment of disease. What kinds of health problems do they deal with? What methods do they use?

3. Visit the library and scan recent science or medical publications for the latest information on a specific disease entity such as AIDS. What is the organism responsible for the disease? Why does it persist in spite of efforts to control it? What is the mechanism through which it produces its characteristic pathology and clinical symptoms? What treatment methods are recommended, and what are the prospects for recovery?

4. Investigate some of the beneficial uses of bacteria, viruses, and protozoa. Identify three or four examples and discuss how these organisms make a positive contribution to human health or society in general.

5. Visit the library and look for recent publications on the current biological classification system. Have there been any changes since R. H. Whitaker devised the original five-kingdom system? If so, what kinds of changes have been made, and why? Are any other changes anticipated?

Challenge Questions

1. Why are Monerans the most metabolically diverse organisms? What is the basis for grouping them in a single kingdom?

2. If viruses are considered nonliving, why are they covered in a course in biology?

3. How are viroids different from viruses? What is a retrovirus?

4. How does the human body respond to a sudden viral invasion?

5. Why aren't we able to develop vaccinations to protect ourselves against certain diseases, such as the common cold?

6. How do antibiotics work? How does an antibiotic such as penicillin kill bacteria?

7. Why do infectious bacteria and other microorganisms tend to thrive in tropical climates?

LESSON 13

Fungi, Plants, and Animals

Assignments

For the most effective study of this lesson, we suggest that you complete the assignments in the sequence listed below:

Before Viewing the Video Program

- Read the Overview and Learning Objectives for this lesson. Use the Learning Objectives to guide your reading, viewing, and thinking.

- Read Chapter 21, "Fungi," pages 332–339; Chapter 22, "Plants," pages 340–355; Chapter 23, "Animals: The Invertebrates," pages 356–381; and Chapter 24, "Animals: The Vertebrates," pages 382–404, in the Starr textbook.

  Or read, in the Starr/Taggart textbook, Chapter 24, "Fungi," pages 388–397; Chapter 25, "Plants," pages 398–415; Chapter 26, "Animals: The Invertebrates," pages 416–445; and Chapter 27, "Animals: The Vertebrates," pages 446–467.

- Read the Viewing Notes in this lesson.

View the "Fungi, Plants, and Animals" Video Program

After Viewing the Video Program

- Briefly note your answers to the questions at the end of the Viewing Notes.

- Review all reading assignments for this lesson, especially the Chapter 21, Chapter 22, Chapter 23, and Chapter 24 Summaries on pages 339, 354, 380, and 403 in the Starr textbook. (Chapter 24, Chapter 25, Chapter 26, and Chapter 27 Summaries on pages 397, 414, 444, and 466 respectively in the Starr/Taggart textbook) and the Viewing Notes in this lesson.

- Write brief answers to all the Review Questions at the end of Chapters 21, 22, 23, and 24 in the Starr textbook (Chapters 24, 25, 26, and 27 in the Starr/Taggart textbook) to be certain you understand the text material.

- Complete the Review Activities in this guide to reinforce your understanding of important terms and concepts. Check your answers with the Answer Key and review when necessary.

- Take the Self-Test in this guide to measure your achievement of the Learning Objectives. Check your answers with the Answer Key and review when necessary.

- Complete the Using What You've Learned activities and any other activities and projects assigned by your instructor.

Overview

In the last lesson, your survey of living organisms took you to the microscopic habitats of bacteria, viruses, and protistans. In this lesson you will leave the world of the single-celled for the world of the multicelled as you explore the evolution of the plants, fungi, and animals that populate our world.

Your journey begins 700 million years ago, when water-dwelling plants were already established in the oceans and plans for establishing a beachhead on land were being slowly implemented. It was cyanobacteria, followed by the green algae and fungi, that first made their way to the banks of freshwater streams and paved the way for the first stalked plants 265 million years later. From then on, plants underwent rapid adaptive radiation, covering much of the land surface within a mere 60 million years. The shift from a haploid to a diploid life cycle, followed by the emergence of pollen grains and seeds, enabled plants to diversify into the various habitats they occupy today.

The evolution of land-dwelling plants is merely a preface to your exploration of the major plant and fungi phyla that populate our world today. Your exploration will include the mosses and liverworts that constitute the bryophytes; the seedless vascular plants such as ferns and horsetails; the conifers, cycads, and ginkgos that make up the gymnosperms; and the flowering, seed-bearing plants known as the angiosperms. You will also examine common features of fungi and learn the reasons for their success after almost a billion years of changing conditions on Earth.

Your journey continues with parallel developments in the animal kingdom, beginning with the emergence of the first multicelled animals from their protistan ancestors 800 million years ago. Sponges and the radially symmetrical jellyfish represent two of the earliest examples. Simple organ systems first appeared with the rise of flatworms and roundworms, followed by the hard-shelled mollusks and segmented worms. Arthropods, which include insects, spiders, centipedes, and the crustaceans, developed a variety of body plan improvements that have made them the most successful organisms on Earth today.

As you continue your journey, you'll find the diversity among invertebrates matched by that of vertebrates, the branch of the animal kingdom where our closest ancestors made their first appearance. From the early chordates and jawless fishes, you'll follow our phylogenetic tree through the more familiar branches—amphibians, reptiles, birds, and mammals—ultimately leading you to the place *Homo sapiens* occupies in the vast scheme of life.

Learning Objectives

When you have completed all assignments for this lesson, you should be able to:

1. Describe the general body plan, modes of nutrition and reproductive patterns of fungi, and distinguish among the groups of *true fungi*.

2. Outline the evolutionary advances that converted ancestors of marine algae into plant forms that could thrive on dry land.

3. Outline the major stages of a plant's life cycle. Identify the cells at each stage as diploid or haploid and whether they are produced by mitosis or meiosis.

4. Identify the *gametophyte* and *sporophyte* stages of algae, mosses, ferns, gymnosperms, and angiosperms. Note the dominant stage in each by comparing the relative amount of time spent in the haploid and diploid states.

5. Compare the general characteristics of the major groups of green plants.

6. Identify the general characteristics of animals.

7. Describe the major evolutionary advances in animal body plans that led to increasingly large and complex organisms.

8. Produce a phylogenetic tree that expresses the evolutionary relationships between the major animal phyla, and describe distinguishing characteristics and example organisms of each.

9. Describe the adaptations that have contributed to the great success of arthropods.

10. Using different classes of vertebrates as examples, provide a general survey of the trends that occurred during vertebrate evolution.

Viewing Notes

Using a lush botanical garden for the appropriate context and inspiration, the video program for this lesson sets the stage for the incredible range of life surveyed in Lesson 13. As Dr. Larry Beezly, curator of Quail Gardens, conducts a visual tour of the kingdom Plantae, Dr. Michael Simpson provides the appropriate interpretation from a biological perspective. Through this survey of plant evolution and the various lineages that have arisen, you'll begin to appreciate how life in general became so diverse and pervasive.

Part One will introduce you to the major plant phyla we see on land today: the leafless liverworts and mosses; the ferns and trees that resulted from the development of a protective cuticle and conductive cells to conserve and move water from one part of the plant to another; the seed plants and flowering plants that evolved new reproductive systems to ensure their dispersion and survival.

In Part Two, you'll continue your survey of the botanical realm with a closer look at the kingdom Fungi. With commercial mushroom grower Bob Crouch, you'll learn about the general characteristics that distinguish more than 100,000 species of fungus from other types of organisms, and you'll develop a healthy respect for one of the most widespread yet unappreciated life forms on the planet. In the process, you'll learn about the important role that fungi play in the ecosystem, including the breakdown of discarded organic material and its symbiotic relationship with plant roots.

In Part Three, you'll join Dr. James Lake and Dr. David Reznick as you leave the world of plants and fungi for the world of animals. With the modern biological classification scheme as your road map, you'll learn of the origins of the first multicelled invertebrates such as sponges, worms, and mollusks. As you progress to more familiar life forms, such as shellfish, insects, and crustaceans, you'll find out about some of the milestones of phylogenetic change. You'll follow the development of the phylum Chordata, beginning with fishes and ending with the land-dwelling vertebrates we know so well today. By the time your survey ends, you will have a greater appreciation not only of the structural differences that characterize individual members of the animal kingdom, but the historical and contemporary linkages that connect each species with every other in the web of life.

As you watch the video program, consider the following questions:

1. How did plants make the transition from water to dry land?
2. What are the major classifications of plants that we see around us today, and how do they differ?
3. Why did flowering plants become the dominant form of plant life on land?
4. Why are fungi among the most prevalent life forms on Earth? What accounts for their success?
5. What are the common structural features of fungi?
6. In what ways do fungi present a danger to humans? In what ways are they helpful?

7. What kinds of evolutionary milestones led to the diversification of animals on the planet?

8. What life forms signaled the emergence of the vertebrates? What kinds of animals were the first vertebrates?

9. In what way are all species on Earth "connected," in spite of their diversity?

Review Activities

Matching

Match the terms below with the definitions that follow. Check your answers with the Answer Key and review any terms you missed.

I.

___ 1. angiosperm
___ 2. bryophyte
___ 3. conifer
___ 4. cycad
___ 5. dicot
___ 6. fern
___ 7. fungus
___ 8. gymnosperm
___ 9. lichen
___ 10. monocot
___ 11. parasite
___ 12. saprobe
___ 13. vascular plant

a. a nonvascular land plant that requires free water to complete its life cycle

b. a type of flowering plant with seeds that have one cotyledon (seed leaf)

c. a type of flowering plant with seeds that have two cotyledons (seed leaves)

d. a seedless vascular plant that lives in wet, humid habitats

e. a type of gymnosperm that is usually woody and evergreen

f. a plant that bears seeds at exposed surfaces of reproductive structures

g. a type of slow-growing gymnosperm that lives in the tropics

h. a plant that has tissue for transporting water and solutes

i. a eukaryotic heterotroph that uses extracellular digestion and absorption

j. a heterotroph that gets its nutrients from nonliving organic matter

k. an organism that gets its nutrients from the tissues of a living host

l. a symbiotic relationship between a fungus and a photosynthetic organism

m. a flowering plant

II.

___ 1. annelid
___ 2. arthropod
___ 3. bivalve
___ 4. cephalopod
___ 5. chordate
___ 6. cnidarian
___ 7. crustacean
___ 8. echinoderm
___ 9. hominid
___ 10. lancelet
___ 11. mollusk
___ 12. rotifer
___ 13. tunicate

a. an invertebrate typically found in food webs of lakes and ponds

b. a radially symmetrical invertebrate that has nematocysts which may cause stings; examples include jellyfish and sea anemones

c. an invertebrate with a tissuefold draped around a soft fleshy body; examples include snails and clams

d. an invertebrate with arms or tentacles strategically arranged around the mouth; examples include squid and octopi

e. a segmented invertebrate; examples include earthworms and leeches

f. an invertebrate with a hardened exoskeleton; examples include spiders and insects

g. a type of invertebrate with calcified spines, needles, or plates on its body wall; examples include sea stars and sea urchins

h. an animal having a notocord, a dorsal hollow nerve cord, a pharynx, and gill slits for part of its life cycle

i. a marine invertebrate with a hard, but flexible exoskeleton; examples include crabs and lobsters

j. an invertebrate chordate that secretes a gelatinous or leathery sac around itself; examples include sea squirts

k. a translucent, fish-shaped invertebrate chordate that lives on the seafloor

l. an invertebrate with a two-part shell; examples include clams and oysters

m. all species on the evolutionary branch leading to modern human beings

III.

___ 1. bilateral symmetry   ___ 6. metamorphosis
___ 2. bipedalism   ___ 7. molting
___ 3. cephalization   ___ 8. radial symmetry
___ 4. exoskeleton   ___ 9. segmentation
___ 5. hydrostatic skeleton

a. the evolutionary process that resulted in the concentration of sensory structures and nerve cells in the head

b. a series of body units that are externally similar or substantially different from one another

c. the shedding of exoskeleton, epidermis, hair, or other body covering as part of growth or renewal

d. habitual standing and walking on two feet

e. an external skeleton, as in arthropods

f. a fluid-filled cavity or cell mass that provides structural support for an organism

g. a body plan having four or more equivalent parts arranged around a central axis

h. a body plan in which the left and right halves of the organism are mirror images

i. major transformation from larva to adult form

Completion

Fill in each blank with the most appropriate term from the list for that paragraph. A term may be used once, more than once, or not at all. If a question requires two or more answers in succession, they may be in any order (unless the question indicates otherwise). Check your answers with the Answer Key and review when necessary.

1. The origin of plant life on Earth began with marine _____ and the subsequent appearance of _____ and _____ on land about 430 million years ago. Adaptation to drier land conditions required the development of _____, _____, and complex mechanisms for _____ and _____. Today, members of the plant kingdom include _____ plants such as bryophytes (which include mosses, liverworts and hornworts); _____, _____ plants called ferns (which include lycophytes and horsetails); _____ plants with exposed _____ called gymnosperms (which include conifers, cycads, and ginkgos); and plants with _____ or angiosperms.

seed(s)	water conservation
seedless	fungi
haploid dominance	algae
flowers	vascular
plants	nonvascular
reproduction	root systems
photosynthesis	monocot
vascular tissues	dicot

2. The multicellular organisms called fungi are _____ and _____. Some feed on living hosts (_____) while others feed off dead organic matter (_____). Fungi form both sexual and asexual _____ which are distinctive for each of the three major fungal classes (_____, _____, _____). _____ are those fungi which cannot be easily classified in a formal taxonomic group.

sac fungi	decomposers
gnetophytes	consumers
zygomycetes	club fungi
parasites	imperfect fungi
lichen	heterotrophs
yeasts	autotrophs
saprobes	hyphae
spores	

3. The origin of animal life is a long and interesting tale. Beginning with ancient multicellular _____ of the Cambrian period, the evolution of animals has been characterized by an incredible diversity of _____, _____, and _____. _____ first appeared with the cnidarians, and _____ followed with the emergence of flatworms and roundworms. An evolutionary split in embryonic development eventually led to the evolution of two lineages: _____, _____ and _____, which are among the most successful animals in the world, and _____ and _____. It was the _____ that began the branch of the phylogenetic tree that eventually led to _____ such as fish, amphibians, reptiles, birds, and mammals. Vertebrate evolution was stimulated not only by the development of a strong _____, but also the appearance of _____ and specialized _____.

vertebrates	body plans
chordates	crustacean
notochords	mollusks
tunicate	organs
sponges	paired fins
tissues	arthropods
respiratory structures	exoskeleton
morphological features	endoskeleton
echinoderms	behaviors
annelids	

Self-Test

Select the one best answer for each question. Check your answers with the Answer Key and review when necessary.

1. All fungi obtain nutrients from
   a. photosynthetic pathways.
   b. nonliving organic matter.
   c. living host organisms.
   d. compounds synthesized by other organisms.

2. True fungi such as sac fungi and club fungi are similar in that
   a. they produce a characteristic type of spore.
   b. they grow in similar habitats.
   c. they have similar body structures.
   d. all of the above are similarities.

3. The evolutionary progression from marine algae to land plants depended on which of the following?
   a. development of underground root structures as anchors
   b. emergence of internal pipelines to transport water and nutrients
   c. origin of an outer cuticle to reduce water loss
   d. all of the above

4. The life cycle of most plants is
   a. dominated by a haploid phase.
   b. dominated by a diploid phase.
   c. equally divided between haploid and diploid phase.
   d. either haploid or diploid, depending on the time of year.

5. Which of the following groups of land plants is the only one to have independent gametophytes with attached, dependent sporophytes?
   a. bryophytes
   b. ferns
   c. angiosperms
   d. gymnosperms

6. The principal difference between gymnosperms and angiosperms is that angiosperms

   a. produce seeds.

   b. produce pollen.

   c. produce flowers.

   d. are vascular plants.

7. Which of the following characteristics is unique to the kingdom Animalia?

   a. they are heterotrophs

   b. they reproduce sexually

   c. they use oxygen for aerobic respiration

   d. none of the above

8. One of the evolutionary developments that led to increases in the size and activity of animals was the emergence of

   a. radial symmetry to provide protection from the environment.

   b. a coelom to cushion and protect vital organs.

   c. segmentation to provide redundant body parts.

   d. respiration to acquire oxygen for metabolic reactions.

9. Differences in embryonic cleavage between protostomes and deuterostomes led to

   a. the evolutionary divergence of mollusks and chordates.

   b. the creation of radially and bilaterally symmetrical body plans.

   c. differences between reptiles and amphibians.

   d. the extinction of the trilobites.

10. Why are the Arthropods considered the most successful organisms on Earth?

    a. They are the oldest members of the animal kingdom.

    b. They occupy more habitats than any other type of organism.

    c. They require less energy to survive and reproduce than any other organism.

    d. All of the above apply.

11. One of the trends that contributed to the evolution of the vertebrates was
    a. the development of a strong internal skeleton that muscles could work against.
    b. the appearance of paired fins, the forerunners of limbs.
    c. the emergence of respiratory structures such as gills.
    d. described by all of the above.

12. The first land vertebrates were
    a. fishes.
    b. amphibians.
    c. reptiles.
    d. mammals.

Using What You've Learned

Based on your own interests or your instructor's requirements, complete one or more of the following activities.

1. Design a flowchart showing the role of fungi in the carbon cycle. Show the flow of energy, exchange of materials, source materials, and products in each phase of the cycle.

2. Construct a chart comparing the life cycle of gymnosperms to that of angiosperms. What are the similarities? What are the differences?

3. Visit the library and find information on the various uses of fungi in medicine and industry. Select one or two specific examples and explain what they are used for. Describe the kind of fungi used, how they are processed, and what makes them especially useful in this capacity.

4. As an alternative, visit the library and research the various uses of *algae* in medicine or the food industry. Again, select one or two specific examples and discuss how they are used in each case.

5. Visit the library to learn more about prehistoric plant life. Find out how, over the course of evolution, natural forces influenced the appearance and prevalence of lycophytes, ferns, and gymnosperms.

6. Using a phylogenetic tree chart, trace the evolution of a particular animal species from its ancestral roots to its present form. Use the chart to explain the morphological similarities between the species and its closest relatives.

7. Investigate what happens when new animal species are abruptly introduced into a foreign habitat. For example, non-native placental mammals introduced into Australian habitats are outcompeting native marsupial populations. In other cases, the native populations continue to dominate. What factors determine which species survives in these situations?

Challenge Questions

1. Why are Protozoans no longer included in the kingdom Animalia?
2. Do mosses really only grow on the north side of trees? If so, why is this?
3. How do fungi such as yeast support fermentation?
4. Why are some fungi not considered "true" fungi? What does this tell you about the biological classification system used today?
5. Why do you think that sponges and cnidarians used to be classified as plants?
6. What is the difference between vertebrates and chordates?
7. Why do populations often collapse when a new predator is introduced into their habitat?
8. How did the muscle cells of birds evolve to accommodate the increased oxygen and energy needs for flight?

LESSON 14

Plants: Tissues, Nutrition, and Transport

Assignments

For the most effective study of this lesson, we suggest that you complete the assignments in the sequence listed below:

Before Viewing the Video Program

- Read the Overview and Learning Objectives for this lesson. Use the Learning Objectives to guide your reading, viewing, and thinking.

- Read Chapter 25, "Plant Tissues," pages 406–421, and Chapter 26, "Plant Nutrition and Transport," pages 422–435, in the Starr textbook.

  Or read, in the Starr/Taggart textbook, Chapter 29, "Plant Tissues," pages 480–499, and Chapter 30, "Plant Nutrition and Transport," pages 500–513.

- Read the Viewing Notes in this lesson.

View the "Plant Structure" Video Program

After Viewing the Video Program

- Briefly note your answers to the questions at the end of the Viewing Notes.

- Review all reading assignments for this lesson, especially the Chapter 25 and Chapter 26 Summaries on pages 420 and 434 in the Starr textbook (Chapter 29 and Chapter 30 Summaries on pages 498 and 512 respectively in the Starr/Taggart textbook) and the Viewing Notes in this lesson.

- Write brief answers to all the Review Questions at the end of Chapters 25 and 26 in the Starr textbook (Chapters 29 and 30 in the Starr/Taggart textbook) to be certain you understand the text material.

- Complete the Review Activities in this guide to reinforce your understanding of important terms and concepts. Check your answers with the Answer Key and review when necessary.

- Take the Self-Test in this guide to measure your achievement of the Learning Objectives. Check your answers with the Answer Key and review when necessary.

- Complete the Using What You've Learned activities and any other activities and projects assigned by your instructor.

Overview

Having been introduced to the various members of the plant kingdom in Lesson 13, Lesson 14 will give you a closer look at the general construction of those plants we are most familiar with—the angiosperms or flowering plants. In the course of studying their basic architecture, you will find out what plants like to eat and how they extract these nutrients from the surrounding environment and distribute them to all parts of the plant.

The lesson begins with a general overview of three plant tissue systems—ground tissues, vascular tissues, and dermal tissues—and the major plant structures that emerge from these tissues. While the video program for Lesson 14 highlights the key features of structures, your textbook provides the details of their composition. Through a combination of line drawings, photographs, time-lapse video and computer animation, you'll gain an understanding and appreciation of the complex tissues and processes that can be found beneath the surface of the seemingly passive roots, stems, and leaves of flowering plants.

One of the first processes you'll learn about is the process of plant growth. You'll learn how the self-perpetuating, embryonic cells of the apical meristems help lengthen the plant (primary growth) while lateral meristems control changes in diameter (secondary growth). These explanations will help you understand common plant growth patterns and phenomena such as tree rings.

Further study of these primary structures will reveal how nutrients essential to plant growth and survival are absorbed and transported within the plant. Your knowledge of chemical bonding and molecular cohesion gained from earlier lessons will help you appreciate how these substances are extracted from the soil and surrounding air, converted into useful compounds, and moved around the plant in defiance of gravitational forces. It will also help you appreciate how plants are able to minimize water loss.

When you've completed Lesson 14, you will have a better idea of not only how plants function, but how form has followed function to produce the diversity of plant life we enjoy—and depend on—today.

Learning Objectives

When you have completed all assignments for this lesson, you should be able to:

1. Describe the generalized body plan of a flowering plant.

2. Define and distinguish among the various types of *ground, vascular,* and *dermal* tissues.

3. Note the locations of *meristems*, explain how plant tissues develop from them, and distinguish *primary* growth from *secondary* growth.

4. Describe the various functions of stems, leaves, and roots.

5. Explain how secondary growth occurs in woody dicot roots and stems, and distinguish *early (spring) wood* from *late (fall) wood*.

6. Identify the nutrients which are essential to plant health, and describe how plant structures obtain these nutrients.

7. Explain how water is absorbed, transported through the *xylem*, and lost by a plant. Note the role of transpiration, osmotic pressure, and the cohesiveness of water.

8. Describe how a plant balances its need for gas exchange with the need to prevent water loss.

9. Explain how the translocation of organic substances occurs in *phloem*, according to the *pressure flow theory*.

10. Describe some of the methods and tools botanists use to study plant physiology.

Viewing Notes

The video program for Lesson 14 will focus your attention on some of the key features that are common to the dominant members of the plant kingdom—flowering plants. From experts in the field you will learn about the general anatomy of flowering plants, how they obtain their nutrients, and how they transport these nutrients to various parts of the plant. Amid the various interviews with botanical experts and explanations of plant tissues and structures, two broad themes will emerge that are central to the program. The first of these is that plants are an essential element in the web of life. Without them, in fact, human life could not exist. Second, plants are far from passive creatures. Under their seemingly motionless exterior lies a virtual beehive of metabolic activities that seem to defy the laws of physics.

Your study of plant anatomy begins literally from the ground up as Part One explores the role of roots in plant survival. Dr. Ann Hirsch, who is studying the roots of plants, explains how roots provide not only an anchor for the plant, but also the principal means for collecting water and nutrients. You'll learn about the special tissues that allow roots to absorb water through osmosis, and the various ways macronutrients such as nitrogen and trace elements are extracted from the surrounding soil. Finally, you'll learn about the implications of these processes to agriculture and our own sustenance.

Part Two of the video breaks ground to explore the stems of plants. More than mere structural supports, you'll learn how the stems of plants provide an expressway for the distribution of water and nutrients to various parts of the plant. Beginning with an explanation of meristems and their role in plant growth, Dr. Arthur Gibson distinguishes between the plant's xylem—the internal pipeline that moves water up the plant *against the force of gravity*—and the *phloem*, the network of sieve tubes that transport sugar from the leaves to the stems and roots. By the end of this segment, you'll have a better appreciation for the range of activity that occurs in plants at the molecular level.

In the third and final portion of the video program, Dr. Robert Heath discusses what is perhaps the most ingenious plant structure of all—leaves. With the help of Dr. Heath and some amazing photomicrography, you'll actually witness the opening and closing of guard cells in the leaf's stomata, which regulate the intake of carbon dioxide and loss of water through transpiration. Dr. Heath will also share his research on the effects of air pollutants such as ozone on plant growth and agricultural productivity.

As you watch the video program, consider the following questions:

1. Why are plants so critical to the survival of animal life on Earth?
2. What are the three basic anatomical structures common to all flowering plants?
3. How do plants such as legumes "fix" nitrogen? How do other plants obtain this nutrient?
4. How do root hairs regulate the intake of water and nutrients?
5. What are meristems, and what is their role in plant growth?

6. How is wood formed?
7. How do tree rings help scientists understand the climatic history of an area?
8. How does the xylem enable the plant to move water against the force of gravity from its roots to its leaves?
9. How do the sieve tubes of the phloem enable the plant to distribute sugar molecules?
10. How do guard cells prevent the excessive loss of water from the plant?
11. What impact do pollutants such as ozone have on plants? What are the broader implications of this?
12. What kinds of technologies can scientists use to study the exchange of energy between plants and their environment?

Review Activities

Matching

Match the terms listed below with the definitions that follow. Check your answers with the Answer Key and review any terms you missed.

____ 1. bud
____ 2. cuticle
____ 3. dermal tissue system
____ 4. fibrous root system
____ 5. ground tissue system
____ 6. leaf
____ 7. meristem
____ 8. nitrogen fixation
____ 9. node
____ 10. parenchyma
____ 11. periderm
____ 12. phloem
____ 13. roots
____ 14. root nodules
____ 15. shoot
____ 16. stoma
____ 17. taproot system
____ 18. translocation
____ 19. transpiration
____ 20. vascular cylinder
____ 21. vascular tissue system
____ 22. xylem

a. a localized region of self-perpetuating, embryonic cells in the stems and roots

b. the most abundant tissue in flowering plants that helps with photosynthesis, storage, secretion, and other functions

c. localized swelling on the roots of certain plants that contain nitrogen-fixing bacteria

d. a protective covering that replaces the epidermis of vascular plants showing secondary growth

e. the lateral branches of adventitious roots in a monocot plant

f. the part of vascular plants that has the chlorophyll-containing tissue critical to photosynthesis

g. the conduction of organic compounds through the body of a vascular plant via the phloem

h. tissue in vascular plants that transports water and solutes through the plant's body

i. an undeveloped shoot consisting primarily of meristem tissue

j. the below-ground parts of vascular plants that absorb and store water and nutrients

Lesson 14 / Plants: Tissues, Nutrition, and Transport   173

k. the point on the stem of a vascular plant where one or more leaves are attached

l. a covering of waxes and cutin that is deposited on the outer surface of the plant's epidermis

m. a primary root and its lateral branchings

n. the combination of xylem and phloem that conducts water and solutes through the body of vascular plants

o. the arrangement of vascular tissues in plant roots

p. the loss of water from stems and leaves by evaporation

q. the gap between guard cells in plant stems and leaves that can be opened or closed to allow the passage of carbon dioxide and water vapor

r. tissues that make up the bulk of the vascular plant body

s. process by which bacteria convert gaseous nitrogen to ammonia

t. the above-ground parts of vascular plants

u. the system of tissues that cover and protect the surface of vascular plants

v. tissue in vascular plants containing the one-way cells that interconnect to form tubes for solute transport

Completion

Fill in each blank with the most appropriate term from the list for that paragraph. A term may be used once, more than once, or not at all. If a question requires two or more answers in succession, they may be in any order (unless the question indicates otherwise). Check your answers with the Answer Key and review when necessary.

1. Flowering plants consist of _____ and _____. _____ include the parenchyma cells that make up the bulk of _____ and participate in _____. _____ include vascular tissues such as _____ and _____, and _____ tissues. The cells of the _____, which are _____ at maturity, provide a network of tubes for conducting _____ and _____. The cells of the _____, which are _____ at maturity, act as storage areas for _____ and other products of _____.

   sugars   respiration
   alive   photosynthesis
   dead   dermal
   carbon dioxide   vascular tissues
   fibrous root system   simple tissues
   ground tissue systems   complex tissues
   dermal tissue system   water
   phloem   nutrients
   meristem   stoma
   cambium   xylem

2. While the _____ of flowering plants contain the cells for photosynthesis, the _____ absorb water and raw materials. Water is conducted through _____ to other parts of the plant including the _____, where water eventually evaporates through open _____. The _____ of water molecules moves the water against the force of gravity. Plant surfaces are covered with a waxy _____ to prevent excessive water loss. According to the pressure-flow theory, organic compounds are _____ through a sieve-tube system where _____ act as the source and _____ serve as the sink.

cuticle   stomata
dermis   roots
electrical cohesion leaves
evaporation   stems
transpiration   meristem
translocated   electrical cohesion
cambium   concentration
pressure   vascular bundles

3. The growth of flowering plants begins at the _____ meristems in the tips of _____ and _____. The embryonic cells of these _____ meristems give rise to the _____ meristem, protoderm, and procambium, which are the principal sources of _____ growth. _____ growth occurs in the _____ cambium and _____ cambium, located in the lateral meristems of the _____ and _____. In _____ plants, seasonal differences produce alternating bands of early and late growth which appear as _____ growth layers.

roots   primary
stems   secondary
shoots   tertiary
stomata   apical
flowering   fibrous
woody   ground
annual   cork
perennial   vascular

Self-Test

Select the one best answer for each question. Check your answers with the Answer Key and review when necessary.

1. Which of the following structures is **NOT** common to all flowering plants?

    a. roots

    b. stems

    c. leaves

    d. fruit

2. Which of the following systems is responsible for conducting water, nutrients, and organic substances throughout the plant body?

    a. ground tissue system

    b. dermal tissue system

    c. vascular tissue system

    d. root tissue system

3. Primary plant growth typically begins at the

    a. apical meristem.

    b. ground meristem.

    c. vascular cambium.

    d. cork cambium.

4. The part of the plant that is responsible for the photosynthetic conversion of sunlight to energy is the

    a. flowers.

    b. leaves.

    c. stems.

    d. roots.

5. Secondary growth in woody dicots often produces

    a. large xylem cells with thin walls (early wood).

    b. small xylem cells with thick walls (late wood).

    c. alternating bands of early and late wood.

    d. either a or b.

6. Which of the following is an essential plant nutrient?

   a. aluminum

   b. sodium

   c. nitrogen

   d. all of the above

7. The primary force that allows water to travel through the xylem against the force of gravity is the

   a. increased pressure due to narrowing xylem conduits.

   b. electrical cohesion among water molecules.

   c. solute concentration differences between the roots and the leaves.

   d. pressure built up at the source of the xylem's sieve-tube system.

8. One way that plants balance the need for carbon dioxide and water conservation is by

   a. ceasing all metabolic processes during the daytime.

   b. keeping stoma closed during the day and open during the night.

   c. secreting a waxy cuticle which is permeable to light.

   d. all of the above.

9. Translocation of organic solutes in the phloem depends on

   a. decreasing concentration gradient from a source to a sink.

   b. increasing pressure buildup in flowers and fruits.

   c. evaporation of water at the plant's stoma.

   d. the wavelike action of the walls of the phloem.

10. Infrared pictures are used by scientists to

    a. determine the composition of plant structures.

    b. follow the transport of nutrients in the xylem.

    c. show primary and secondary growth patterns.

    d. monitor temperature changes at the stoma.

Using What You've Learned

Based on your own interests or your instructor's requirements, complete one or more of the following activities.

1. Using pictures of leaves or actual leaf specimens, explain how leaf morphology varies with climate and lighting conditions. To what extent do the leaves provide protection or camouflage from predatory insects or other animals? What are some of the possible adaptive advantages of the different sizes, shapes, and colors exhibited?

2. Find five or six examples of unusual plant tissues (e.g., cotton bolls, nutshells, hemp fibers) and discuss what kinds of tissues they represent. Explain in each case why these particular structures evolved.

3. Design a hypothetical plant that would function best under each of the conditions listed below. Sketch the plant, showing the size and shape of basic structures such as roots, stems, and leaves. Explain how processes such as nitrogen and CO_2 fixation, photosynthesis, nutrient transport, water conservation, and plant growth might be different under these conditions:

   Condition A: High sunlight, high temperature, and extreme drought
   Condition B: Low sunlight, constant rain, hot and humid climate
   Condition C: High altitude, cold temperature, and minimal atmosphere

4. Visit the library and find out about useful chemical products that are made from plants (e.g., insecticides, medications). Select one product in particular and explain how it is extracted from the plant and modified. What are its uses? How prevalent is the source plant? Is it cultivated? Or is it in danger of becoming extinct?

5. Visit a lumberyard and find out what kinds of trees are the best source of lumber. Obtain samples and describe their suggested uses.

6. Visit a local nursery and investigate the special requirements for growing plants in your local area. Do conditions favor drought-resistant plants? Or is there an abundance of water? What kinds of soils predominate, and what kinds of plants grow best in them? Determine which plants are best suited to the area and what kinds of care (e.g., fertilizers, irrigation) are recommended. Explain the rationale for these recommendations based on what you know about the anatomy and physiology of plants in general.

7. Investigate the kinds of soils that are best for supporting plant growth. How do fertilizers augment the soil? How are soils tested to determine their composition?

8. Find out about the ways in which plants combat insect pests in nature. What man-made methods are used to help plants in this process, and which appear to be most effective?

9. Investigate the science of *dendrochronology*—using tree growth patterns to study the natural history of an area. Explain how growth rings are used to reveal information about normal growth, past droughts, fires, loss of trees due to windfall, disease, harvesting, etc. Find a cross-section of a tree and try to interpret the significance of its growth rings.

Challenge Questions

1. Why do plants need nitrogen to survive? What kinds of molecules depend on nitrogen, and what would happen if these molecules were unavailable?
2. Why is it important to include some of the surrounding soil when moving a plant from one place to another?
3. What would happen if stomata lost their ability to open and close? What would happen if they were permanently open? What would happen if they were permanently closed?
4. Can plants be overwatered? Why is this bad for the plant?
5. Why is it important to avoid overplanting an area? What do you think happens when too many plants compete for nutrients in the same soil?
6. How do the tissues found in the xylem conduct water even though they are dead at maturity?
7. How does the growth of plants differ from that of animals?
8. What is tree sap? What is it made of, and what is its purpose?
9. What is the source of cork, the substance used to make bottle stoppers? Why is this substance uniquely suited for this purpose?

LESSON 15

Plants: Reproduction and Development

Assignments

For the most effective study of this lesson, we suggest that you complete the assignments in the sequence listed below:

Before Viewing the Video Program

- Read the Overview and Learning Objectives for this lesson. Use the Learning Objectives to guide your reading, viewing, and thinking.

- Read Chapter 27, "Plant Reproduction and Development," pages 436–456, and review Chapter 21, "Fungi," pages 332–339, and Chapter 22, "Plants," pages 340–355, in the Starr textbook.

  Or read, in the Starr/Taggart textbook, Chapter 31, "Plant Reproduction," pages 514–527, and Chapter 32, "Plant Growth and Development," pages 528–542. Also review Chapter 24, "Fungi," pages 388–397, and Chapter 25, "Plants," pages 398–415.

- Read the Viewing Notes in this lesson.

View the "Plant Reproduction" Video Program

After Viewing the Video Program

- Briefly note your answers to the questions at the end of the Viewing Notes.

- Review all reading assignments for this lesson, especially the Chapter 27 Summary on page 455 in the Starr textbook (Chapter 31 and Chapter 32 Summaries on page 526 and 541 respectively in the Starr/Taggart textbook) and the Viewing Notes in this lesson.

- Write brief answers to all the Review Questions at the end of Chapter 27 in the Starr textbook (Chapters 31 and 32 in the Starr/Taggart textbook) to be certain you understand the text material.

- Complete the Review Activities in this guide to reinforce your understanding of important terms and concepts. Check your answers with the Answer Key and review when necessary.

- Take the Self-Test in this guide to measure your achievement of the Learning Objectives. Check your answers with the Answer Key and review when necessary.

- Complete the Using What You've Learned activities and any other activities and projects assigned by your instructor.

Overview

A brief review of the major classification of plants and their distinctive life cycles will provide all the background you need to delve into the biochemical processes, cellular events, and environmental interactions that characterize sexual reproduction in plants, the theme of Lesson 15. The lesson will enhance your appreciation for the methods behind the madness of colors, patterns shapes, and aromas that characterize flowering plants. It may also give you a new perspective on the role of *Homo sapiens* and other species in the perpetuation of plants all over the world.

Your study of plant reproduction begins with a survey of the anatomy and physiology of flowering plants, including the site of male and female gamete production. Unlike vertebrate animals, plants are an all-in-one reproduction machine containing both male and female sexual organs as well as all the necessary equipment for embryonic packaging. Yet while plants contain the means for asexual reproduction, most prefer sexual reproduction for the adaptive advantages it provides. So plants have evolved a wide range of strategies for ensuring the mating of male sperm cells or spores with the egg cells of different members of the population. This is the process of pollination, a process that depends on the forces of nature or symbiotic relationships with other organisms to provide the mobility which plants themselves lack.

Pollination is but one of the fascinating processes you'll learn about in Lesson 15. You'll also learn about what happens after fertilization produces a new embryo sporophyte and eventually a seed. You'll learn about the structures and events that package the seed for dispersal and support it prior to germination. You'll learn how that seed matures into an adult plant capable of participating in the ritual of reproduction. And you'll learn of internal and external factors that influence growth of plants and the timing of life-cycle events.

As you explore these processes, Lesson 15 will introduce you to some of the work being done to understand plant reproduction and growth. You'll learn of botanists and horticulturists who intervene in these processes to protect endangered species and perpetuate those plants that play a crucial role in our survival. You'll also learn how humans have learned to manipulate the chemical and physical environment of plants to accelerate or impede their growth.

Learning Objectives

When you have completed all assignments for this lesson, you should be able to:

1. Draw and label the parts of a *perfect* flower. Explain how gamete formation occurs in the male and female structures.

2. Define and distinguish between *pollination* and *fertilization*. Explain how the coevolution of flowering plants and certain animals has aided the pollination process.

3. Describe the components of a seed and list various methods of seed dispersal, including the role of fruit. Explain how the coevolution of some plants and animals has also aided in the dispersal process.

4. Identify the factors that cause plants to *germinate*, and describe the general pattern of early plant growth.

5. List the various *hormones* that regulate plant growth and metabolism, and describe their known effects on plants.

6. Explain how plants respond to stimuli or changes in their environment, and describe the factors that might cause a plant to flower, age, or enter *dormancy*.

7. Describe the techniques botanists use to investigate plant reproduction and development.

Viewing Notes

By watching the video program for Lesson 15, you'll begin to appreciate why plants are among the "busiest organisms on Earth." Far from idle, the plant's machinery for growth and reproduction are constantly in motion, often involving other members of the biosphere in its struggle for survival.

Without a means for moving from one place to another, each type of plant has devised an ingenious strategy to reproduce with other plants and spread its offspring to other habitats. In Part One of the program, you'll see some of these strategies and what really happens within the familiar structures we associate with members of the plant kingdom. You'll hear from various botanists as they describe the evolutionary rationale for these structures and their functions. As in other programs, these descriptions will be enhanced by computer animation and time-lapse photography that manipulates time and space to show you what happens during reproductive processes such as gamete formation, fertilization, and pollination.

To learn about the symbiotic relationships between plants and insect pollinators such as the bee, you'll meet beekeeper Alan Mikolich, who explains how apiaries such as his aid in the pollination of different crops. This kind of relationship is of particular interest to Dr. Travis Columbus, who studies the reproductive mechanisms of plants in an effort to understand why some plant species become endangered. In Part Two, Dr. Columbus tells "A Seedy Story," in which he explains the structure and development of seeds and their dispersal through a wide variety of strategies. One of these strategies is the production of edible fruit that will be consumed, digested, and eliminated by other organisms, depositing the surviving seeds in a new location for eventual germination and growth.

While explaining the germination process, Dr. Columbus introduces the importance of environmental cues such as moisture, heat, and light. These same environmental cues also play a role in mature plant development, as you'll see in Part Three. In this segment, Dr. Elliot Meyerowitz will also introduce you to some of the internal cues or hormones that influence plant development. You'll learn about the role of gibberellins in seed germination and stem growth. You'll learn about the effect of ethylene on the seedling's orientation to gravity and auxins in the maturing plant's response to gravity and other environmental stimuli. And you'll find out how seasonal changes and other environmental cues affect the life cycle of plants.

As you watch the video program, consider the following questions:

1. The program calls plants some of the "busiest organisms on Earth." What is meant by this statement?
2. What is the primary function of flowers?
3. What parts of flowering plants are responsible for gamete production?
4. What is pollination, and why is it important to the survival of plant species?
5. What kinds of strategies have evolved among plants to ensure seed dispersal?

6. Why did natural selection favor the evolution of edible fruit? What purpose do they serve?

7. How does the structure of a typical seed contribute to its growth and survival?

8. What kinds of environmental cues cause a seed to germinate?

9. What are the classes of hormones that influence plant growth and development, and what are their functions?

10. What are some examples of "tropism," and what purpose does it serve in each case?

Review Activities

Matching

Match the flower structures pictured below with the terms that follow. Some terms may be used more than once. Check your answers with the Answer Key and review any terms you missed.

I.

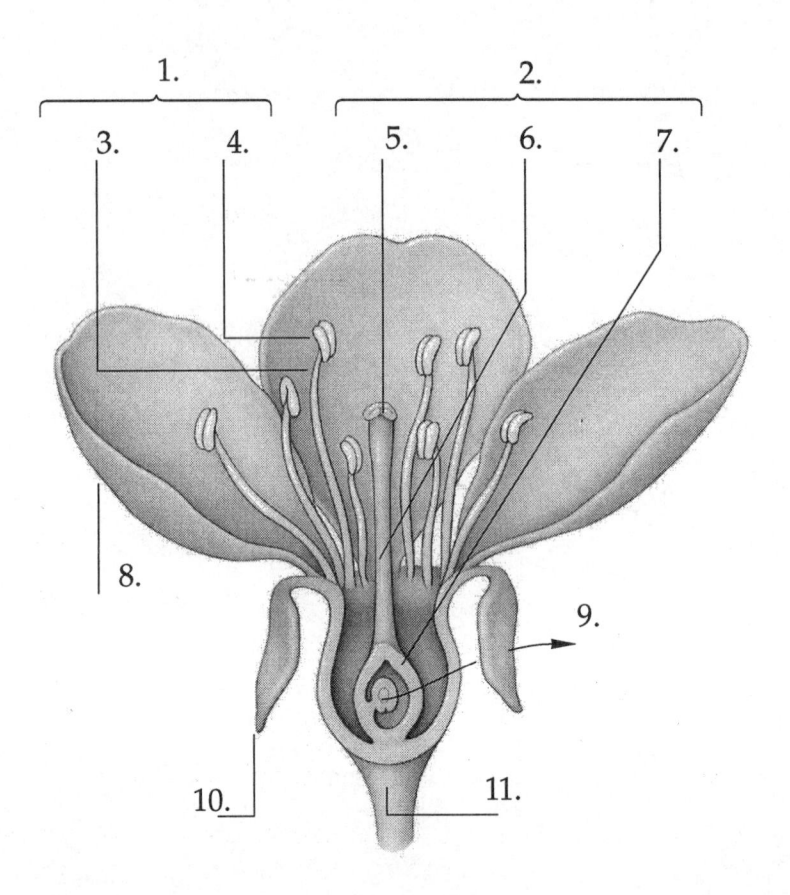

a. ovary
b. stamen
c. carpel
d. style
e. petal
f. receptacle
g. filament
h. stigma
i. sepal
j. anther
k. ovule

Lesson 15 / Plants: Reproduction and Development

Now match the terms below with the definitions that follow. Check your answers with the Answer Key and review any terms you missed.

II.

____ 1. auxin
____ 2. dormancy
____ 3. fertilization
____ 4. gametophyte
____ 5. germination
____ 6. gravitropism
____ 7. megaspore
____ 8. microspore
____ 9. phototropism
____ 10. pollination
____ 11. propagation
____ 12. photoperiodism
____ 13. phytochrome
____ 14. senescence
____ 15. sporophyte
____ 16. thigmotropism

a. the fusion of an egg nucleus and a sperm nucleus to form a zygote

b. tendency of a plant to adjust its direction of growth in response to gravity

c. the arrival of a pollen grain on the stigma of a flowering plant

d. light-sensitive pigment molecule that triggers hormonal-controlled growth activities when activated

e. a haploid spore that forms in the ovary and develops into an egg

f. a growth-regulating hormone in plants

g. aging processes leading to the natural death of all or part of an organism

h. the tendency of a plant to adjust its direction and rate of growth in response to light

i. a vegetative body that grows from a zygote and produces the plant's spore-bearing structures

j. the haploid, gamete-producing phase in the plant's life cycle

k. biological response to changes in the relative length of daylight and darkness

l. a temporary, hormone-induced interruption of growth under conditions that would normally promote growth

m. a haploid spore encased in a sculpted wall that develops into a pollen grain

n. tendency of a plant to adjust its direction of growth in response to physical contact with solid objects

o. the point at which an embryo sporophyte breaks through its seed coat and continues to grow

p. the process of causing a plant or other organism to multiply by natural reproduction

Completion

Fill in each blank with the most appropriate term from the list for that paragraph. A term may be used once, more than once, or not at all. If a question requires two or more answers in succession, they may be in any order (unless the question indicates otherwise). Check your answers with the Answer Key and review when necessary.

1. Flowering plants usually reproduce _____. The _____ produces _____ where haploid spores develop into _____. Male _____ become sperm-carrying _____ while female _____ containing eggs form in the _____. A pollen grain landing on a receptive _____ develops a pollen tube that brings sperm into contact with the egg for _____. Once fertilized, the newly formed _____ develops into a mature _____ sporophyte. As the _____ grows, it stores nutrients from the parent plant into the expanding _____ or leaflike structures called _____. Eventually the _____ containing the embryo develops a hard coat and separates from the _____ wall. The resulting _____ is dispersed as part of the plant's _____ and may take root and grow into an adult plant.

ovule	zygote
ovary	sexually
pollination	fertilization
pollen grains	flowers
sporophyte	fruit
stamen	embryo
megaspore	stigma
endosperm	seed
coleoptile	gametophyte(s)
cotyledons	germination

2. Plants also reproduce _____ through a variety of mechanisms. _____ produces an embryo from a fertilized egg. The root system of some plants can give rise to separate _____. Strawberry plants send out _____ that yield new _____ and _____ at alternate nodes. New _____ can also arise from axillary _____ on _____ and the underground _____ of other kinds of plants. Plants can be artificially cloned through the process of _____.

roots	buds
shoots	bulb
sexually	stems
asexually	flower
parthenogenesis	fruit
induced propagation	tubers
shoot systems	runners
rhizome	

3. The life cycle of plants is influenced by a variety of factors. Hormones such as _____ and _____ promote stem growth, while _____ promote leaf expansion. Abscisic acid triggers bud and seed _____, and _____ helps fruits ripen. Hormones also influence the _____ and _____ of growth on each side of the plant so it can make _____ responses to _____, _____, and _____. _____ plants respond to changes in the relative length of daylight and darkness. Even the plant's own _____ produces rhythmic leaf movements on a daily basis.

light	cytokinins
temperature	gibberellins
gravity	biological clock
photoperiodic	quality
phytochrome	type
auxins	direction
ethylene	rate
benzene	senescence
dormancy	tropic

Lesson 15 / Plants: Reproduction and Development

Self-Test

Select the one best answer for each question. Check your answers with the Answer Key and review when necessary.

1. Which of the following flower parts is the source of male sperm-producing pollen grains?

    a. stigma

    b. sepal

    c. carpel

    d. anther

2. In a flowering plant, egg production occurs in the

    a. anther.

    b. ovary.

    c. stigma.

    d. style.

3. Which of the reproductive processes depends most on external forces such as air currents or animal behaviors?

    a. germination

    b. fertilization

    c. pollination

    d. propagation

4. Which of the following best describes the function of a cotyledon?

    a. It is the part of the plant that gives rise to a developing embryo.

    b. It protects the plant embryo from the external environment.

    c. It is the growing edge of a developing plant embryo.

    d. It provides nourishment for a developing plant embryo.

5. Which of the following is **NOT** a significant factor in germination?

    a. gravity

    b. amount of daylight

    c. soil temperature

    d. amount of oxygen in the soil

6. Which of the following plant hormones is most directly responsible for the abscission of leaves, flowers, and fruits?

   a. abscisic acid

   b. auxins

   c. ethylene

   d. gibberellins

7. Plants do not usually become dormant until

   a. they have completed at least one reproductive cycle.

   b. the winter solstice reduces daylight hours to a minimum.

   c. external temperatures drop below a specific level.

   d. all appropriate environmental changes have occurred.

8. Induced propagation is often used by horticulturists to

   a. produce clones of plants from shoot cuttings.

   b. promote pollination when natural mechanisms are ineffective.

   c. generate plant embryos from unfertilized eggs.

   d. simulate conditions necessary for a seed to germinate.

Using What You've Learned

Based on your own interests or your instructor's requirements, complete one or more of the following activities.

1. Visit a botanical garden or local nursery. Observe the various structures that plants have evolved to promote fertilization and pollination. Construct a chart with photos or diagrams comparing and contrasting these structures.

2. Bees are not the only insects that play a major role in the plant life cycle. Identify at least three other insects or animals that help plants grow and reproduce, and describe their role in detail.

3. Visit the library and scan recent science publications on the treatment of hay fever and other pollen-related allergies. What kinds of treatments are most effective?

4. Using the library, research the use of hydroponic techniques to raise plants under artificial conditions. How are lighting and soil conditions simulated? How are seasonal cues provided?

5. Use the library to survey the various methods agriculturists use to inhibit the growth of harmful plants or propagate the growth of desirable plants. Which techniques are most effective?

6. Select a particular form of vegetation and use a map to show where it grows best. Explain why conditions favor the survival of this particular plant over others and how it might have spread to occupy its current habitat(s).

7. Based on your knowledge of plant growth and tropisms, speculate on how plant life would evolve on other planets with the basic elements to sustain growth. For example, how would plants evolve on a planet with lower or higher gravity? Colder or hotter surface temperatures? Lower or higher light intensity? Shorter or longer periods of daylight?

Challenge Questions

1. What features had to evolve to enable plants to flourish on land?

2. What strategies for pollination are most efficient? What strategies for seed dispersal are most efficient?

3. What are the similarities between a pollen grain and a human sperm cell? What are the differences?

4. If edible fruits provide an adaptive advantage for seed dispersal, why are some fruits inedible or even poisonous?

5. Why do different kinds of plants grow at different elevations? What factors related to altitude affect plant growth and development?

6. How do herbicides work? Why do they kill some plants and not others?

7. How are plant hormones used by commercial growers to manipulate plant growth?

8. What is the difference between a fruit and a vegetable?

9. How do plants with seedless fruits (grapes, watermelons, etc.) propagate?

LESSON
16

Animals: Structure and Movement

Assignments

For the most effective study of this lesson, we suggest that you complete the assignments in the sequence listed below:

Before Viewing the Video Program

- Read the Overview and Learning Objectives for this lesson. Use the Learning Objectives to guide your reading, viewing, and thinking.

- Read Chapter 28, "Tissues, Organ Systems, and Homeostasis," pages 458–471, and Chapter 32, "Protection, Support, and Movement," pages 524–539, in the Starr textbook.

  Or read, in the Starr/Taggart textbook, Chapter 33, "Tissues, Organ Systems, and Homeostasis" pages 544–557, and Chapter 38, "Protection, Support, and Movement," pages 626–647.

- Read the Viewing Notes in this lesson.

View the "Animal Structure" Video Program
After Viewing the Video Program

- Briefly note your answers to the questions at the end of the Viewing Notes.

- Review all reading assignments for this lesson, especially the Chapter 28 and Chapter 32 Summaries on pages 470 and 538 in the Starr textbook (Chapter 33 and Chapter 38 Summaries on pages 556 and 646 respectively in the Starr/Taggart textbook) and the Viewing Notes in this lesson.

- Write brief answers to all the Review Questions at the end of Chapters 28 and 32 in the Starr textbook (Chapter 33 and Chapter 38 in the Starr/Taggart textbook) to be certain you understand the text material.

- Complete the Review Activities in this guide to reinforce your understanding of important terms and concepts. Check your answers with the Answer Key and review when necessary.

- Take the Self-Test in this guide to measure your achievement of the Learning Objectives. Check your answers with the Answer Key and review when necessary.

- Complete the Using What You've Learned activities and any other activities and projects assigned by your instructor.

Overview

Lesson 16 marks your transition from the world of plants to the world of animals. Beginning with a look at meerkats, tortoises, and other species, you will explore the relationship between form (anatomy) and function (physiology) in the animal kingdom.

The lesson begins by defining the basic anatomical building blocks—tissues, organs, and organ systems—and describing examples of each. You will learn about epithelial tissue and its role in establishing boundaries between body compartments and protecting the body's internal environment from outside assaults. This is the function of the skin, the largest living organ in the human body, and the principal component of the integumentary system.

You will also learn about connective tissue and the various forms it takes. In addition to providing general support for other tissues and organs, specialized connective tissues in the form of bone, blood, and fat perform important homeostatic functions. Bone tissue, for example, plays a crucial role in blood cell formation and calcium exchange.

In the course of your study, you will learn about nerve and muscle tissue, which interact to produce the muscle contractions necessary for movement. With your prior knowledge of metabolic reactions, you will be able to better appreciate the intricate biochemical processes that cause voluntary muscle cells to contract and relax on demand. Your understanding of these processes will also serve as a useful foundation for learning how muscles, bones, and joints interact to support the organism and allow it to move.

As you complete this lesson, think about the ultimate purpose of the various tissues, organs, and organ systems: to ensure the stability of the organism's internal environment. Just as the internal chemical balance of the individual cell is crucial for its survival, so too is the internal chemical balance of the entire organism. Maintaining the stability of this internal environment—a condition known as *homeostasis*—is the driving force behind the design and function of the unique anatomical structures that distinguish one animal species from another. Your study of homeostasis and its implications for survival will only begin in Lesson 16; it will continue over the next several lessons as you explore the major organ systems that characterize members of the animal kingdom.

Learning Objectives

When you have completed all assignments for this lesson, you should be able to:

1. Define the terms *tissue*, *organ*, and *organ system*, and use them to describe the general structure of complex animals.

2. Describe the basic structure and function of the four types of animal tissues. Identify the types of cells that compose each tissue, and describe the location and function of each tissue in example organs.

3. Define *homeostasis*, and describe how *feedback control mechanisms* help maintain this condition.

4. Describe the structure of the human *integumentary system*, and discuss the skin's role in maintaining homeostasis.

5. Identify the general functions of a vertebrate's *skeletal system*, and describe the basic structure of a typical long bone.

6. Describe the structure of a freely moving *joint*—noting the role of *cartilage*, *tendons*, and *ligaments*—and describe some disorders associated with joints.

7. Compare "skeletal" and "muscular" components of earthworms, crayfish, and humans, and describe how these components interact to produce movement.

8. Describe the composition and structure of muscles, and explain how muscle cells contract using the *sliding-filament model*.

9. Summarize the steps that lead to muscle contraction, noting the roles of *motor neurons*, calcium ions, and ATP molecules.

10. Compare the effects of aerobic and anaerobic activity in causing muscle soreness and fatigue, and describe how regular exercise can help muscles resist fatigue.

Viewing Notes

In the video program for Lesson 16, you will gain a greater appreciation of the diversity within the animal kingdom and how organisms with radically different physical structures have found their own solutions to the problem of maintaining homeostasis. The adaptive advantage of these structures within the organism's natural environment is made evident by observing what happens when animals such as the desert tortoise are relocated to an entirely different setting.

With Professors David Reznick and Vaughn Shoemaker, you will descend through the protective layers of the skin to explore the composition of the internal environment, with its various tissues, organs, and organ systems. You will discover why the skin itself is considered an organ and learn about its role in water retention, temperature regulation, and other homeostatic processes.

In Part Two, "Muscles in Microgravity," Professor Kenneth Baldwin and Dr. Joan Vernikos will take you into space for a look at the effect of weightlessness on human physiology and describe the difficulties astronauts encounter when they try to readapt to normal Earth gravity. By studying the loss of muscle mass and strength that affects returning astronauts, scientists hope to learn more about Earth-bound problems such as the decrease in muscle mass which naturally occurs as we age. In the process, you'll learn about the neurochemical mechanisms responsible for muscle contraction.

In Part Three of the program, "Living Levers," you'll visit Old English Rancho to see fine examples of thoroughbred horse, and the kinds of physical attributes that make for a winning racehorse. This segment provides an appropriate introduction to the role of the skeletal system in determining the shape and mobility of an animal. Through this discussion, you will learn about the development and composition of bone tissue and its role in maintaining homeostasis. You will also learn about joints and how they provide a fulcrum for the musculoskeletal lever systems that are found in the arms and legs of animals. As in other lessons, you'll see how form and function go hand in hand to produce the incredible diversity that characterizes the animal kingdom.

As you watch the video program, consider the following questions:

1. What is the "internal environment," and why is its stability important to survival?
2. What are the roles of tissues, organs, and organ systems in maintaining homeostasis?
3. What are the different kinds of tissues found in animals, and what are their functions?
4. How does the integument of an animal help it survive in its surroundings?
5. Why do astronauts living in a prolonged weightless state experience muscle weakness? How can this be remedied?
6. What role does calcium play in maintaining the integrity and function of the muscular and skeletal systems?
7. How do muscles contract?

8. How does regular exercise build muscle strength and endurance?
9. How are the survival needs of different animals reflected in the structure of their skeletal and muscular systems?

Review Activities

Matching

Match the terms listed below with the definitions that follow. Check your answers with the Answer Key and review any terms you missed.

I.

_____ 1. adipose tissue
_____ 2. cartilage
_____ 3. connective tissue
_____ 4. epithelium
_____ 5. integument
_____ 6. muscle tissue
_____ 7. nervous tissue
_____ 8. organ
_____ 9. organ system
_____ 10. tissue

a. animal tissues that contain fibroblasts and other cells which secrete protein fibers and polysaccharides

b. in multicelled organisms, a group of cells and intercellular substances that work together to perform one or more specialized tasks

c. tissue that is made up of neurons and neuroglia

d. two or more organs that interact physically and/or chemically

e. tissue that is made up primarily of fat-storing cells

f. tissue consisting of cells that have the ability to contract when stimulated

g. a well-defined anatomical structure that is composed of more than one tissue

h. animal tissue made up of adhering cells that covers the body's external surfaces and lines its internal cavities

i. a type of connective tissue made up of solid yet pliable intracellular material that resists compression

j. in animals, an organ system that provides a protective body covering

II.

___ 1. effector
___ 2. endocrine
___ 3. exocrine
___ 4. homeostasis
___ 5. integrator
___ 6. negative feedback mechanism
___ 7. positive feedback mechanism
___ 8. sensory receptors

a. a process whereby a change in the environment triggers a response that compounds (or increases) the change

b. a ductless gland that secretes hormones into interstitial fluid

c. a muscle or gland that produces movement (or chemical change) in response to changing environmental conditions

d. a control point where information is collected and actions selected in response to changing environmental conditions

e. a process whereby a change in the environment triggers a response that reverses (or decreases) the change

f. gland that secretes hormones to a free epithelial surface through ducts or tubes

g. nerve cells that are designed to detect a specific stimuli, then relay the signal to the spinal cord or brain

h. in multicellular organisms, a state in which the internal chemical and physical conditions are maintained within acceptable limits

III.

____ 1. actin   ____ 5. myofibril
____ 2. action potential   ____ 6. tendon
____ 3. dermis   ____ 7. sarcomere
____ 4. ligament   ____ 8. sarcoplasmic reticulum

a. the membrane system that stores or recycles calcium ions required for muscle contraction

b. a bundle of dense connective tissue that bridges two or more bones at a joint

c. a bundle of dense connective tissue that attaches muscle to bone

d. the basic unit of contraction in vertebrate muscles

e. a globular protein that interacts with another protein (myosin) during muscle contraction

f. a threadlike structure inside a muscle cell that contains the basic units of contraction

g. a brief reversal in the otherwise steady voltage difference across the plasma membrane of nerve cells

h. a layer of dense connective tissue that lies just below the epidermis

Completion

Fill in each blank with the most appropriate term from the list for that paragraph. A term may be used once, more than once, or not at all. If a question requires two or more answers in succession, they may be in any order (unless the question indicates otherwise). Check your answers with the Answer Key and review when necessary.

1. Complex animals have developed specialized groups of _____ called _____ which are organized into more well-defined structures called _____. When _____ interact, they comprise a(n) _____. The survival of any given animal depends on how effective these structures are in maintaining the stability of the body's _____, a condition known as _____. The body uses positive and negative _____ to ensure that changes in its internal environment are reversed before they get out of control.

cells	integrators
effectors	integument
electrical charges	internal environment
endoderm	molecules
epithelium	organs
exoskeleton	organ systems
feedback mechanisms	physiology
homeostasis	tissues

Lesson 16 / Animals: Structure and Movement

2. Most animals are made up of four kinds of tissues: _____, _____, _____, and _____. _____ has a free surface and is useful in lining internal and external body structures. Examples of epithelial tissue can be found in the _____ and _____ or _____ glands. _____ is composed of _____ and is designed to conduct _____. _____ consists of cells that _____ when stimulated and _____ when relaxed. _____ supports other body structures, or performs more specialized functions as in the case of _____, _____, _____, or _____.

adipose tissue	epithelial tissue
blood	exocrine
bone	expand
cartilage	fibroblasts
charged particles	integument
collagen	lengthen
connective tissue	mesoderm
contract	muscle tissue
electrical impulses	nervous tissue
endocrine	neurons

3. In animals, three organ systems are responsible for the body's protection, shape, and movement: the _____, the _____, and the _____. The _____ in vertebrates consists of an outer layer of stratified _____ (_____) and an underlying layer of _____ (_____), which includes muscle and nervous tissue. The skeletal system consists of _____, which are connected at the _____ by _____, _____, or _____. The bones provide _____, _____, and a mechanism for the exchange of _____ used in metabolism. Muscles are attached to bone by _____. A typical muscle is composed of threadlike _____ that are divided into _____, the basic unit of contraction. Through _____, the _____ and _____ filaments that make up the _____ interact and slide past each other, contracting the muscle fibers. This interaction is the main premise of the _____ of muscle contraction.

actin
action potential model
blood cell(s)
bone(s)
cartilage
connective tissue
creatine phosphate
cross-bridge formation
dead
dermis
epidermis
epithelial cell(s)
fibrous tissue
keratinocytes
joint(s)
ligaments

living
melanin
mineral ion(s)
movement
motor unit
myofibril(s)
myosin
muscle(s)
protection
sarcomere(s)
sarcoplasmic reticulum
shape
sliding-filament model
support
tendon(s)

Self-Test

Select the one best answer for each question. Check your answers with the Answer Key and review when necessary.

1. Which of the following would be considered an organ?
   a. blood
   b. skin
   c. epithelium
   d. skeleton

2. Connective tissue(s)
   a. includes bone and cartilage.
   b. transmit electrical impulses to and from sensory receptors.
   c. have at least one free surface.
   d. are primarily used to connect muscle to bone.

3. Homeostasis is best defined as a state in which
   a. the organism's internal environment is in balance.
   b. organs and organ systems "shut off" when not in use.
   c. the destruction of old tissue matches the production of new tissue.
   d. feedback mechanisms are no longer needed to regulate metabolism.

4. Which of the following is **NOT** a function of the integumentary system?
   a. regulation of body temperature
   b. protection against ultraviolet radiation
   c. destruction of harmful bacteria
   d. maintenance of body shape and movement

5. The typical mature long bone consists primarily of
   a. white marrow.
   b. osteoblasts.
   c. compact bone tissue.
   d. cartilage.

6. Osteoarthritis is a disorder that results from
    a. wearing away of the cartilage covering the bone ends of joints.
    b. the gradual replacement of cartilage with bone deposits.
    c. inflammation and thickening of the synovial joints.
    d. bacterial or viral infection at the joints.

7. The skeleton of a crayfish differs from that of a human in that the crayfish skeleton
    a. does not allow movement.
    b. redistributes body fluid within a limited space.
    c. is composed of flexible rather than rigid parts.
    d. is external rather than internal.

8. Skeletal muscle cells consist of
    a. myofibrils bundled in parallel arrays.
    b. neurons encased in smooth, connective tissue.
    c. free-floating sarcomeres that line up when stimulated.
    d. a gridlike pattern of myosin and actin molecules.

9. The principal mechanism that enables muscles to contract is the
    a. depletion of calcium ions at the sarcomeres.
    b. electrical charge of the sarcoplasmic reticulum.
    c. interaction of actin and myosin filaments.
    d. buildup of creatine phosphate in the muscle cells.

10. Regular aerobic exercise can help muscles resist fatigue by
    a. increasing the size of mitochondria in the muscle cells.
    b. improving the blood supply to fast and slow muscle cells.
    c. stimulating the growth of additional muscle tissue.
    d. reducing dependence on fatty acids for fuel.

Using What You've Learned

Based on your own interests or your instructor's requirements, complete one or more of the following activities.

1. Identify examples of positive and negative feedback mechanisms in daily life that do not occur in living organisms. Draw a diagram showing how each works, and relate these examples to similar homeostatic processes in the human body.

2. Using microscope slides or pictures of such slides, sketch and label each of the four types of tissues. Explain how function follows form in each case.

3. Obtain an uncooked whole or half chicken from the supermarket. Examine each of the systems discussed in the lesson, starting with the integument. Can you identify the epidermis and dermis? Next inspect the muscle tissue, noting how it is attached to the bird's skeleton. Where are the largest muscles located? Cut into a muscle, noting the density and consistency of the tissue. Cut away some of the muscles so that you can examine the skeleton. Compare and contrast how different joints are constructed. Look for examples of tendons, ligaments, and cartilage, comparing the characteristics of these tissues and how they attach to muscle and bone.

4. Cut an ordinary bone such as a chicken bone lengthwise and examine its contents. What kinds of structures do you see? Is it a young bone or a mature bone?

5. Visit the library and scan recent science and/or health publications for current research on any of the following disorders and related treatment:

    - skin cancer (including the risks posed by tanning beds or sunlight)
    - spinal disorders such as spina bifida, scoliosis, lordosis, kyphosis
    - sprains, strains, dislocations and bone fractures
    - muscle diseases such as myasthenia gravis and muscular dystrophy
    - bone and joint diseases such as osteoporosis and arthritis

    Select an interesting article on one or more of these disorders and prepare a brief report summarizing the principal outcomes.

6. Identify and research a particular exercise approach or regimen. Explain how it would contribute to strength, flexibility, endurance, and/or overall vitality based on your knowledge of the muscular and skeletal systems.

7. Construct models of the three principal types of joints—fibrous, cartilaginous, and synovial—to illustrate how each works. Cite examples of each type of joint in the human body.

Challenge Questions

1. What kinds of metabolic problems would you expect if homeostatic mechanisms suddenly became inoperative?
2. How does the body's regulation of glucose provide an example of a homeostatic mechanism? Is it a positive or negative feedback mechanism?
3. How does the structure of epithelial tissue, connective tissue, nerve tissue or muscle tissue provide insight into its function in the body?
4. Why is blood considered a connective tissue? Why is a bone considered an organ?
5. What is the role of ATP in muscle contraction?
6. How is muscular contraction coordinated to produce smooth movement?
7. Why do we have different kinds of joints? How does construction of a given joint reflect its function?

LESSON 17

Animals: Circulation

Assignments

For the most effective study of this lesson, we suggest that you complete the assignments in the sequence listed below:

Before Viewing the Video Program
- Read the Overview and Learning Objectives for this lesson. Use the Learning Objectives to guide your reading, viewing, and thinking.

- Read Chapter 33, "Circulation," pages 540–561, in the Starr textbook.
  Or read, in the Starr/Taggart textbook, Chapter 39, "Circulation," pages 648–669.

- Read the Viewing Notes in this lesson.

View the "Circulation: A River of Life" Video Program
After Viewing the Video Program
- Briefly note your answers to the questions at the end of the Viewing Notes.

- Review all reading assignments for this lesson, especially the Chapter 33 Summary on page 560 in the Starr textbook (Chapter 39 Summary on page 668 in the Starr/Taggart textbook) and the Viewing Notes in this lesson.

- Write brief answers to all the Review Questions at the end of Chapter 33 in the Starr textbook (Chapter 39 in the Starr/Taggart textbook) to be certain you understand the text material.

- Complete the Review Activities in this guide to reinforce your understanding of important terms and concepts. Check your answers with the Answer Key and review when necessary.

- Take the Self-Test in this guide to measure your achievement of the Learning Objectives. Check your answers with the Answer Key and review when necessary.

- Complete the Using What You've Learned activities and any other activities and projects assigned by your instructor.

Overview

Complex organisms that depend on oxygen to sustain cell growth and reproduction must have a way to take oxygen in from the external environment and deliver it to the cells. In vertebrates, the respiratory system facilitates oxygen intake; transport is handled by the blood and circulatory system.

Lesson 17 examines this crucial system and its various components. Beginning with the evolutionary origins of vertebrate circulation, the lesson uses morphological comparisons to show how this system is configured in different species within the animal kingdom. You'll learn about the differences between open and closed circulatory systems and the advantages and limitations of each.

In a highway system where the vehicles have little mobility of their own, movement must be provided by the system itself. In closed circulatory systems, this movement is provided by the heart pump and a system of interconnected blood vessels. The ability of the smaller vessels to expand and contract as needed ensures a steady flow of blood to the capillaries and to the cells beyond. Lesson 17 will explore these and other mechanisms for regulating blood pressure and blood flow.

The lesson also examines the principal medium of the circulatory system: the blood. Part of the internal chemical environment that bathes and nourishes each cell of our body, the blood is a remarkably versatile substance. In addition to containing the red cells that act as the medium of exchange for oxygen and carbon dioxide, the blood is home to the white cells necessary for infection control. The blood also transports nutrients absorbed from the digestive tract to the cells and carries away waste products for excretion through the skin, kidneys and lungs. It plays a major role in temperature regulation and provides the substances necessary for tissue repair.

Lesson 17 will include a brief survey of blood disorders, and you will see for yourself the kinds of consequences that develop when circulating levels of any of the basic blood elements are disrupted. Other circulatory disorders affecting the heart and blood vessels will also be discussed. Learning about these disorders will not only help you better understand the normal function of the circulatory system, but also appreciate the tremendous role that medicine plays in restoring and maintaining cardiovascular health.

The lesson concludes with a look at a separate but closely related system, the lymphatic system, which drains excess blood plasma and delivers absorbed body fats to the bloodstream. The lymphatic system, which includes the spleen and thymus gland, also plays a major role in cleansing the body tissues of pathogens and cellular debris.

Learning Objectives

When you have completed all assignments for this lesson, you should be able to:

1. Describe the general function of a *circulatory system*, and identify its three principal components.

2. Compare *open* and *closed* circulatory systems with respect to structure and operation. Use different organisms as examples.

3. Describe the components of human blood, and list the basic functions of each.

4. Use the human heart to trace the path of blood flow through the *pulmonary* and *systemic circuits*, and compare this pathway with that of other vertebrates.

5. Describe the basic structure of the human heart, explain its operation through a complete *cardiac cycle*, and identify the heart's mechanisms of contraction.

6. Compare the structure and function of *arteries, arterioles, capillaries, venules* and *veins*. Note the direction of blood flow through each and the degree of nutrient/waste exchange occurring through the walls.

7. Define the term *blood pressure*, and describe the mechanisms that regulate it.

8. Describe common disorders of the *cardiovascular system*, and discuss some possible causes of those disorders.

9. Outline the basic structure and function of the *lymphatic system*, and explain how it works with the cardiovascular system to recover *interstitial fluids*.

10. Describe how scientists examine cardiovascular systems.

Viewing Notes

The video program for Lesson 17 is aptly entitled "Circulation: A River of Life." From its very beginning, the program establishes the importance of the circulatory system by describing the clinical consequences when the system is compromised by traumatic injury. Like a river unable to supply life-sustaining water and nutrients to the plants and animals along its banks, the bloodstream is unable to deliver vital oxygen and nutrients to the cells. Within minutes, the cells begin to die and vital organs cease to function.

Dr. Barry Heller uses his life and death experiences in the operating room as a springboard for describing the normal function of the human circulatory system. The system is a simple one, consisting of three primary components: 1) the heart, the small but tenacious muscle that pumps over a thousand gallons of blood a day; 2) the intricate system of blood vessels that serves as the main superhighway for blood; and 3) the blood itself. Through videomicrography and computer simulation, you'll get a firsthand look at each of these components and how they interact to constitute the "river of life."

In the course of exploring the anatomy of the circulatory system and the physics of blood flow, you'll learn something about blood pressure and its role in maintaining circulation. You'll learn about the factors that influence blood pressure and what it means when blood pressure drops below safe limits. And you'll learn about the body's built-in safety measures to restore blood pressure when total circulating volume falls to dangerous levels.

With a basic understanding of how the human circulatory system works, you'll be ready to look at the circulatory systems of other animals and how they compare to ours in terms of structure and function. In Part Two, Dr. James Hicks will show you how the blood cells of turtles and alligators differ from human blood cells and what adaptive advantage these cells may provide. He will also contrast the reptilian three-chambered heart with our own four-chambered heart and discuss the possible reasons for this and other genetic adaptations.

Part Three returns to the world of humans to explore the pathology of heart attacks, which continue to be the major cause of death in the United States. You'll follow the case of Jerry Medway, heart attack victim, as he describes his experience and the events following his attack. You'll also hear from Dr. Rahimtoola and Dr. Merz as they provide a medical analysis of Jerry's condition and the steps that were taken to restore his health. In the process, you'll learn about the physiological conditions that precipitate a heart attack, including weaknesses in the heart muscle itself and disruptions in heartbeat called arrhythmias. The program ends with a look at the lifestyle changes Jerry made to avoid future heart problems.

As you watch the video program, consider the following questions:

1. Why is damage to the heart or circulatory system considered life-threatening?
2. What are the three components of the circulatory system?
3. How is blood pumped and recycled throughout the human body?
4. How does the construction of the arterioles, venules, and capillaries facilitate blood flow?
5. What mechanisms control blood pressure? How can blood pressure be artificially restored in emergency situations?
6. How does the blood of animals such as reptiles differ from the blood of humans in terms of oxygen transport?
7. What is the structure of the human heart, and how does it differ from the hearts of other vertebrates?
8. What is the mechanism that regulates the pumping action of the heart?
9. What is a heart attack, and how does it occur?

Review Activities

Matching

Match the terms listed below with the definitions that follow. Check your answers with the Answer Key and review any terms you missed.

I.

___ 1. agglutination
___ 2. blood pressure
___ 3. cardiac cycle
___ 4. circulatory system
___ 5. hemostasis
___ 6. lymphatic system
___ 7. pulmonary circuit
___ 8. systemic circuit
___ 9. vasoconstriction
___ 10. vasodilation

a. interruption of blood loss from a damaged vessel through natural mechanisms

b. the alternate muscle contraction and relaxation that characterizes one heartbeat

c. an increase in blood vessel diameter caused by relaxation of smooth muscle in the vessel walls

d. a decrease in blood vessel diameter caused by contraction of smooth muscle in the vessel walls

e. blood circulation leading to and from the cells in the body

f. the force generated by heart contractions that keeps blood circulating

g. blood circulation leading to and from the lungs

h. the clumping together of foreign cells caused by cross-linking between attached antibody molecules

i. the system that picks up fluid from the interstitial spaces and cleanses it before delivery to the bloodstream

j. the system responsible for transporting materials to and from the cells

II.

____ 1. interstitial fluid
____ 2. lymph node
____ 3. plasma
____ 4. platelet
____ 5. red blood cell
____ 6. spleen
____ 7. stem cell
____ 8. thymus
____ 9. white blood cell

a. an unspecialized cell that produces daughter cells that divide and differentiate into specialized cells

b. an oxygen-transporting cell

c. a lymphoid organ which acts as a filtering system for blood and a repository for red blood cells

d. a gland containing lymphoid tissue where lymphocytes multiply, differentiate, and mature

e. the portion of extracellular fluid that occupies the space between cells

f. a cell fragment in blood that releases substances used in clot formation

g. one of numerous, pea-shaped organs that contain the lymphocytes and macrophages of the immune system

h. the fluid component of blood

i. among these are the macrophages, eosinophils, and neutrophils that perform immune system functions

III.

___ 1. aorta   ___ 6. capillary bed
___ 2. arteriole   ___ 7. vein
___ 3. artery   ___ 8. ventricle
___ 4. atrium   ___ 9. venule
___ 5. capillary

a. any small blood vessel that transfers blood from capillaries to veins

b. one of the chambers of the heart which receives blood

c. a zone with numerous capillaries where blood and interstitial fluid exchange nutrients and waste materials

d. any thin-walled blood vessel where gases and other substances are exchanged between blood and interstitial fluid

e. any large-diameter blood vessel that conducts blood away from the heart

f. any large-diameter blood vessel that conducts blood back to the heart

g. one of the chambers of the heart that pumps blood out

h. any small blood vessel that transfers blood from arteries to capillaries

i. the largest artery of the circulatory system

Completion

Fill in each blank with the most appropriate term from the list for that paragraph. A term may be used once, more than once, or not at all. If a question requires two or more answers in succession, they may be in any order (unless the question indicates otherwise). Check your answers with the Answer Key and review when necessary.

1. The human circulatory system transports substances to and from cells through a system of _____, _____, and _____. The transport medium is the _____, which consists of _____, _____, _____, and other substances in a liquid _____ solvent. The force needed to circulate the blood is provided by the rhythmic pumping action of the _____. Contraction of the right _____ drives oxygen-poor blood through the _____, where it picks up oxygen. Oxygen-rich blood entering the left _____ is pumped out through the left _____ to the _____ circuit, where gases and other substances are exchanged with the _____. Oxygen-poor blood returns to the heart to the right _____.

   red blood cells   life
   white blood cells   systemic
   interstitial fluid   platelets
   heart   blood
   ventricle   arteries
   plasma   veins
   capillaries   atrium
   pulmonary circuit

2. The _____ is a network of _____ and vessels that acts as a filtration system for _____. _____ and _____ that have leaked from blood at the _____ are returned to the bloodstream along with _____ absorbed from the small intestine. _____ and cellular debris are sent to the _____ for disposal.

   interstitial fluid   pathogens
   red cells   spleen
   white cells   water
   fats   thyroid
   lymph nodes   capillary beds
   lymphatic system   proteins
   thymus

216   Lesson 17 / Animals: Circulation

3. In the bloodstream, red blood cells or _____ transport oxygen and carbon dioxide. White blood cells or _____ perform housekeeping tasks and defend against foreign invaders. _____ are cell fragments that release substances critical to blood clotting.

Blood cells arise from stem cells in the bone marrow. When _____ production fails to meet the body's needs, _____ occurs. _____ occurs when a gene mutation produces red cells with an abnormal form of hemoglobin. _____ results when cancer suppresses the formation of white blood cells. Disorders can also arise from too many blood cells. _____ (too many red cells) and infectious mononucleosis (too many white cells) are two examples.

leukemia	hemolysis
anemia	melanoma
gamma globulin(s)	sickle-cell anemia
polycythemia	erythrocyte(s)
platelet(s)	neurocyte(s)
leukocyte(s)	thrombocytopenia

Self-Test

Select the one best answer for each question. Check your answers with the Answer Key and review when necessary.

1. All of the following are primary components of the circulatory system **EXCEPT**
   a. the heart.
   b. the blood vessels.
   c. the lungs.
   d. blood.

2. Unlike closed circulatory systems, the open circulatory system of insects and mollusks
   a. does not have a heart.
   b. circulates blood in two directions rather than one.
   c. includes a single complete circuit of arteries and veins.
   d. pumps blood into the spaces between body tissues.

3. Which of the following blood components helps the body defend against foreign invaders?
   a. white blood cells
   b. red blood cells
   c. platelets
   d. plasma

4. In the human circulatory system, blood entering the pulmonary circuit for oxygenation is pumped out of the
   a. right atrium.
   b. left atrium.
   c. right ventricle.
   d. left ventricle.

5. The heart is able to perform its pumping function through the
   a. natural pressure gradient between the venous and arterial system.
   b. rhythmic electrical stimulation of cardiac muscle.
   c. chemical changes brought about by the accumulation of deoxygenated hemoglobin.
   d. the "ripple effect" caused by alternating vasoconstriction and vasodilation.

6. Although capillaries are smaller than arterioles, blood flows into them because
    a. the solute concentration of capillary plasma is higher.
    b. the solute concentration of capillary plasma is lower.
    c. arteriol pressure is lower than capillary pressure.
    d. their combined diameter provides less total resistance to blood flow.

7. Blood pressure is determined by all of the following **EXCEPT**
    a. the number of blood cells circulating in the bloodstream.
    b. nervous stimulation of heart muscle.
    c. nervous stimulation of smooth muscle in vessel walls.
    d. hormonal stimulation of smooth muscle in vessel walls.

8. Hypertension can be caused by
    a. plaque or a clot lodged in a blood vessel.
    b. a buildup of cholesterol and other lipids in arterial walls.
    c. an abnormal increase in circulating red cells.
    d. an abnormal decrease in circulating red cells.

9. The primary purpose of the lymphatic system is to
    a. provide a reservoir of additional blood plasma when needed.
    b. manufacture the cellular components of the bloodstream.
    c. cleanse the interstitial fluid of pathogens and debris.
    d. provide a backup system for circulation when the normal circulatory system malfunctions.

10. Electrocardiograms are useful in
    a. mapping the location of key veins and arteries.
    b. locating thromi and emboli that can block blood flow.
    c. blood typing.
    d. identifying irregularities in heart rhythm.

Using What You've Learned

Based on your own interests or your instructor's requirements, complete one or more of the following activities.

1. Using a chart of the circulatory system, trace the route of a typical deoxygenated red cell from the time it leaves the heart to the time it returns to the heart in the same deoxygenated state.

2. Visit the library and scan current science publications for recent research on heart disease. What is the latest thinking in the scientific and medical communities about the causes of heart disease? What behaviors promote it? What behaviors inhibit it? What are the most effective forms of treatment?

3. Visit your family physician and ask to see your electrocardiogram from a recent physical. Ask the physician to interpret the chart, and explain what the readings indicate in terms of the condition of your heart and your health in general.

4. Ask your family physician to explain the use of a stethoscope and, if possible, let you use it to listen to the heart of a volunteer. What does the sound you hear represent? What kinds of sounds does the physician listen for, and what do they mean?

5. Ask the same physician if you can observe the use of a sphygmomanometer for blood pressure measurement. How does it work? What do the systolic and diastolic readings indicate about a person's blood pressure?

6. Find out what your blood type is and what it means in terms of your ability to donate or receive blood transfusions. Based on your study of Lesson 17, what does the blood type say about the composition of your blood?

7. Visit your library or arrange with local law enforcement agencies to research the use of blood typing in criminal investigations. How do criminologists match blood found at the scene of a crime with the blood of a potential suspect? How reliable and accurate are these techniques?

8. Devise a chart showing basic blood elements and their function. Explain what happens when there is a significant excess or deficiency in each of these elements.

Challenge Questions

1. Why is it that people who die from heart attacks have adequate amounts of oxygenated blood in their heart chambers?

2. What does it mean when the heart is referred to as a "double pump"?

3. Although the lymphatic system is supposed to cleanse the blood of harmful substances, it often plays a major role in the spread of cancer. Why is this?

4. What is it that differentiates the blood of a hemophiliac from that of a non-hemophiliac?

5. Why is it that animals with open circulatory systems can function normally when blood flows freely into the interstitial spaces, but animals with closed circulatory systems would consider this kind of internal bleeding a life-threatening condition?

6. How does clotting occur? What happens when clots break off and enter the bloodstream?

7. Before giving an injection, the doctor or nurse always makes sure that there are no oxygen bubbles in the hypodermic. What would happen if oxygen bubbles were injected into the bloodstream?

LESSON 18

Animals: Immunity

Assignments

For the most effective study of this lesson, we suggest that you complete the assignments in the sequence listed below:

Before Viewing the Video Program

- Read the Overview and Learning Objectives for this lesson. Use the Learning Objectives to guide your reading, viewing, and thinking.

- Read Chapter 34, "Immunity," pages 562–581 in the Starr textbook.
  Or read, in the Starr/Taggart textbook, Chapter 40, "Immunity," pages 670–689.

- Read the Viewing Notes in this lesson.

View the "Immunity" Video Program

After Viewing the Video Program

- Briefly note your answers to the questions at the end of the Viewing Notes.

- Review all reading assignments for this lesson, especially the Chapter 34 Summary on page 580 in the Starr textbook (Chapter 40 Summary on page 688 in the Starr/Taggart textbook) and the Viewing Notes in this lesson.

- Write brief answers to all the Review Questions at the end of Chapter 34 in the Starr textbook (Chapter 40 in the Starr/Taggart textbook) to be certain you understand the text material.

- Complete the Review Activities in this guide to reinforce your understanding of important terms and concepts. Check your answers with the Answer Key and review when necessary.

- Take the Self-Test in this guide to measure your achievement of the Learning Objectives. Check your answers with the Answer Key and review when necessary.

- Complete the Using What You've Learned activities and any other activities and projects assigned by your instructor.

Overview

Every animal faces the challenge of defending itself against invasion by other organisms or antigenic substances. In Lesson 18, you will be introduced to an array of defenses against a wide variety of pathogens.

Usually, the body begins its defense by setting up surface barriers to any potential invader. Skin, mucous membranes, protective fluids, and excretory functions are examples of such barriers. In a symbiotic way, nonpathogenic populations of organisms on the surface also act to protect against invasion. The next set of defenses is mostly nonspecific internal responses to any threat. These include the inflammatory actions of white blood cells, macrophages, and substances such as complement proteins. Internal organs like the spleen and lymph nodes contribute to these defenses. Finally, specific invaders are targeted by specific immune responses. These can be rather complex interactions involving white blood cells such as T cells, B cells, and macrophages. Weapons used in this fight include antibodies, perforins, interleukins, and enzymes. All of these defenses are discussed in detail and effectively presented in your textbook and video materials.

As you complete Lesson 18, you will find that even this amazing set of defenses may not adequately protect an individual from the threat of disease. Biologists have attempted to enhance and supplement the system using techniques such as vaccinations, immunotherapy, and desensitization programs. Despite all the efforts of nature and science, a complete defense against the human immunodeficiency virus (HIV) has yet to be found. The resulting disease, widely known as AIDS, is proof that more work needs to be done before we truly understand how to regulate the immune response.

Since this is a complex subject, you will be encouraged to build your knowledge in a stepwise manner, learning the basic immune responses before studying the underlying mechanisms. Your efforts will be rewarded with a greater appreciation of our ability to survive in a very dangerous world!

Learning Objectives

When you have completed all assignments for this lesson, you should be able to:

1. Describe typical surface barriers that vertebrates present to defend against invading *pathogens*.

2. Summarize the process of *acute inflammation*, describe its signs, and explain how this mechanism fights pathogens.

3. Generally compare the body's *nonspecific* and *specific* immune response, noting the basic features which distinguish the latter.

4. Summarize how vertebrates (especially mammals) recognize and discriminate between self and nonself tissues.

5. Describe and distinguish between *cell-mediated* and *antibody-mediated* (humoral) immune response, noting the roles of the different white blood cells and any resulting antibodies.

6. Explain how the lymphatic system's fluid recovery function helps fight infectious agents.

7. Distinguish between the human body's *primary* and *secondary* immune response, noting the role of *macrophages* and various *lymphocytes*.

8. Distinguish between *active* and *passive* immunity, and identify examples of each in modern medical therapy.

9. Describe some examples of abnormal or deficient immune responses in humans, and identify which weapons in the immunity arsenal failed in each case.

10. Describe methods scientists use to study immunity.

Viewing Notes

The video program for Lesson 18 uses several scenarios to illustrate the important features of immune responses. The first of these scenarios will take you to the Orange County Bird of Prey Center, where veterinarian Dr. Scott Weldy explains how he treats and rehabilitates injured birds. His explanations provide a springboard for a discussion of how organisms in general defend themselves against invasion by dangerous pathogens, including the role of the skin, mucous membranes, and finally the cells and antibodies that make up the inflammatory response.

Through this presentation, you will learn about the various players in the drama of inflammation, including the supporting role of the lymphatic system. You'll discover the biological rationale for inflammation, its classic signs and symptoms, and how it relates to healthy immune system function. You'll learn about the roles of red and white blood cells in immune system function and how they contribute to the inflammatory response.

As Part One ends, you will have a better understanding of the body's nonspecific response to infection. But this is only part of the story. Part Two expands the discussion of immunity by describing another line of defense involving cell-mediated and antibody-mediated responses to *specific* invaders such as the HIV virus responsible for AIDS.

For Dr. Eric Galpin, the subject of viral infections and immune system responses is more than academic. Having nearly succumbed to polio in the 1950s, Dr. Galpin now works to find a cure for HIV. Through his guidance and the magic of computer animation, you'll be able to visualize the complex interaction of antigens, antibodies, and cells which constitute the third line of defense of our immune system. In the process, you'll learn where in the body these defenders are manufactured and how they are "programmed" to recognize and destroy foreign invaders without harming our own host cells. Part Two concludes with a look at the significance of the specific immune response in combating public health hazards such as AIDS.

The final portion of the video considers what happens when immune systems go haywire, as they do in autoimmune disorders such as lupus. By following the case of Carolyn, who suffers from lupus, we get a highly personal perspective to counterpoint the clinician's view of autoimmune disorders. In this case, the clinician's perspective is provided by Dr. Robin Dore, who theorizes what happens on a biochemical level when autoimmune disease strikes a host. While doctors are unable to prevent autoimmune disorders or cure people like Carolyn, they are able to relieve the symptoms through medication and other forms of treatment.

As you watch the video program, consider the following questions:

1. What kinds of barriers do we have against invasion by pathogens?
2. What are the signs and symptoms of inflammation, and what are the reasons for them?
3. How do the specific and nonspecific responses to infection differ?

4. Why is the lymphatic system so important in defending against infectious pathogens?
5. Are immune responses different in various parts of the body?
6. What is the biochemical basis for cell-mediated immunity? For antibody-mediated immunity?
7. How do animals respond to a pathogen during the second exposure to its antigens?
8. Can the response to a given pathogen cause more damage than the pathogen itself? If so, how?
9. What theories have been proposed to explain why autoimmune disorders such as lupus occur?
10. What kinds of biochemical and cellular dysfunctions characterize an autoimmune disorder?
11. What are some of the methods and techniques used to study the immune system?

Review Activities

Matching

Match the terms listed below with the definitions that follow. Check your answers with the Answer Key and review any terms you missed.

I.

___ 1. pathogen
___ 2. antigen
___ 3. antibodies
___ 4. complement system
___ 5. effector cells
___ 6. acute inflammation
___ 7. perforins
___ 8. lymphoid organs
___ 9. T lymphocytes
___ 10. B lymphocytes

a. a series of plasma protein reactions capable of causing inflammation, direct lysis of pathogens and stimulation of phagocytosis

b. response to tissue damage and irritation which can include white blood cells, complement proteins and macrophages

c. any disease-causing agent; some are infectious like bacteria and viruses

d. cells originating in bone marrow and maturing in the thymus to acquire TCRs

e. proteins produced by cytotoxic T cells aimed at antigens on target cells

f. any agent or molecule recognized as "nonself" and capable of initiating an immune response

g. cells involved in destroying antigen-containing targets, such as T and B cells

h. specialized tissues and organs including lymph nodes, the spleen, thymus, and others involved in the immune response

i. protein effector molecules manufactured by B cells in response to specific antigenic stimuli

j. cells originating in bone marrow and capable of producing antibodies

II.

___ 1. cytotoxic T cells
___ 2. NK cells
___ 3. helper T cells
___ 4. plasma cells
___ 5. memory cells
___ 6. macrophages
___ 7. MHC markers
___ 8. interleukin
___ 9. primary immune response
___ 10. secondary immune response

a. cells capable of destroying antigenic targets without the formation of antigen-MHC complexes using a "touch-kill mechanism"

b. protein markers characterized as "self," located on cells and used to form complexes with antigens

c. clones of B cells derived from an immune response to specific antigenic stimuli

d. T cells capable of destroying antigenic targets recognized by their antigen-MHC complexes

e. molecules secreted by macrophages and helper T cells to enhance the T cell-mediated immune response

f. T cells capable of communicating with macrophages and other T and B cells; stimulate the cloning of effector and memory cells

g. response to initial encounters with antigens including white blood cells, inflammation and cell or antibody-mediated actions

h. T and B cells formed to respond at a later time to the same antigenic stimulus, important in the secondary immune response

i. a rapid response of memory cells and others, most times to a previously encountered antigen

j. cells capable of destroying antigenic targets by phagocytosis and joining in on T and B cell-mediated responses

III.

___ 1. IgG
___ 2. IgA
___ 3. IgM
___ 4. IgE
___ 5. IgD
___ 6. autoimmune response
___ 7. passive immunization
___ 8. active immunization
___ 9. HIV
___ 10. AIDS

a. immunoglobulins initially secreted, which signal complement reactions and agglutinate for easier phagocytosis

b. immune response directed at one's own cells or antigenic targets defined as "self-markers"

c. immunoglobulins responding to allergens and parasitic antigens, which promote inflammatory response

d. injections of antibodies providing temporary defenses against pathogens already infecting or soon to be in contact with

e. a disease syndrome stemming from an HIV infection

f. immunoglobulins that may be present with IgM on virgin B cells; elevated in chronic infections

g. immunoglobulins capable of activating complement proteins and specific antigen binding on many bacteria, viruses, and parasites

h. virus infecting the immune cell; capable of causing the disease syndrome AIDS

i. immunoglobulins found in mucous membranes and other secretions, capable of neutralizing many antigenic targets; protect newborns through mother's milk

j. injections or oral doses of antigenic compounds capable of producing a primary immune response or a subsequent secondary response

Completion

Fill in each blank with the most appropriate term from the list for that paragraph. A term may be used once, more than once, or not at all. If a question requires two or more answers in succession, they may be in any order (unless the question indicates otherwise). Check your answers with the Answer Key and review when necessary.

1. Inflammatory responses occur when tissues are damaged physically, by _____ or if contacted by a variety of _____ materials. White blood cells including _____, basophils, and _____ enter to protect the tissue from pathogens and further damage. _____ release histamine to promote _____ to bring in more help from the bloodstream. Monocytes differentiate into _____ which destroy many pathogens and help induce fevers to kill most invaders. The _____ proteins also help to stop the invasion while _____ and repair mechanisms attempt to return the area to normal. Some of the signs of _____ include redness, warmth, _____ and pain. The overall strategy is to _____ the damage and promote _____.

    clotting   monocytes
    complement   muscle cells
    contain   neurons
    eosinophils   neutrophils
    healing   palpitations
    infection   swelling
    inflammation   toxic
    macrophages   vasoconstriction
    mast cells   vasodilation

2. The structure of antibodies holds many of the clues behind their specificity. The antibodies are comprised of _____ polypeptide _____ usually presenting a _____. There are two sets of chains called heavy and _____. Each of these is further divided into _____ and variable regions. The variable regions contain the _____ sites specific for the _____ of the invading pathogen. During the maturation process of _____, unique areas of the cells' DNA, called _____ and _____ segments, are changed. These segments combine in a random fashion to form a rearranged _____ sequence. When this virgin B cell encounters a corresponding antigen complex, the B cell is induced to _____. Thus the DNA coding for binding that specific antigen is passed on to each B cell _____.

A	divide	RNA
antigen	DNA	T cells
B	four	two
B cells	J	V
chains	light	X shape
clone	receptor	Y shape
constant		

3. Immune responses triggered by allergens are known as _____. This response can include sinus congestion, eye irritation, _____, hives, increased _____ secretions, and other reactions. While a _____ predisposition to allergies has been shown, responses may depend on the current _____ conditions as well as the physical and mental health of the individual. There are many potential _____ such as airborne particulates and _____, foods, plants, drugs, and biological toxins. Most of the symptoms involve _____ antibodies which bind to the antigen and _____ cells locally present. The mast cells then release _____, leading to inflammation and problems in the respiratory system. Once established, allergic reactions can be severe, even leading to _____ shock and death. While avoidance of the allergen is the preferred defense, minor symptoms can be controlled by _____, and treatments including _____ may help in the long run.

allergens	genetic	insulin
allergies	histamines	mast
anaphylactic	hormone	mucous
antihistamines	IgE	pollens
desensitization	IgM	revascularization
environmental	inflammation	T cells

Lesson 18 / Animals: Immunity 231

Self-Test

Select the one best answer for each question. Check your answers with the Answer Key and review when necessary.

1. All of the following are surface barriers to pathogens EXCEPT
   a. skin.
   b. saliva.
   c. antibodies.
   d. sufficient symbiotic bacteria or "normal flora."

2. Acute inflammation is characterized by all of the following EXCEPT
   a. interleukin secretion.
   b. macrophage phagocytosis.
   c. activation of the complement system.
   d. None of the above is characteristic of acute inflammation.

3. Specific immune responses are characterized by
   a. inflammation.
   b. specificity of antigens or pathogens.
   c. fever production.
   d. basophil and eosinophil activity.

4. Which of the following allows for recognition of nonself antigens?
   a. cells presenting antigen-MHC complexes
   b. mucous membrane barriers
   c. survival of organisms after a fever
   d. detection of mRNA in the cytoplasm

5. Antibody-mediated immune responses include
   a. production of specific B-cell clones.
   b. production of specific perforins.
   c. production of NK cells.
   d. production of antihistamines.

6. Which of the following lymphoid organs defend against infection by packing their tissues with antigen-presenting cells and other lymphocytes?

   a. tonsils
   b. spleen
   c. lymph nodes
   d. all of the above

7. Secondary immune responses are

   a. initiated by memory cells.
   b. initiated by virgin cells.
   c. initiated by first encounters with pathogens.
   d. far slower and weaker than primary immune responses.

8. The main difference between passive and active immunity is

   a. active immunity employs antibodies for protection.
   b. passive immunity uses only IgG antibodies.
   c. active immunity can only be induced by killed pathogens.
   d. passive immunity is limited and temporary.

9. Rheumatoid arthritis is an example of an immune system failure known as

   a. SCID.
   b. anaphylactic shock.
   c. autoimmune response.
   d. immunological memory.

10. One method used by research scientists to enhance the immune response against cancer is

    a. developing effective LAK cells.
    b. developing effective monoclonal antibodies.
    c. developing effective vaccines.
    d. described by all of the above.

Using What You've Learned

Based on your own interests or your instructor's requirements, complete one or more of the following activities.

1. Describe at least ten sources of allergies that might be found in your home. Discuss various ways the potential health effects of each allergen could be minimized.

2. Visit your local library for references in diagnostic microbiology. List the bacteria considered to be "normal flora" in the human mouth and throat. Are some of these also pathogens? How do you explain their continuing presence in those areas?

3. There are several ways to screen out microbes in commercial beer. Name at least two brands using pasteurization and two using a type of filtration. Call the consumer line of each and identify yourself as a biology student. Ask them to describe methods used to prevent contamination of their products.

4. Diagram the immune response to breathing in spores of a pathogenic fungus.

5. Obtain free information about childhood vaccinations from your local health department clinic. Do you think the benefits outweigh the risks for these immunizations?

6. Obtain the most recent cumulative statistics on HIV infection, those diagnosed with AIDS, and the resulting mortality figures. Do these figures fit the definition of an epidemic? If not, can you suggest a reason?

7. Read about the history of smallpox and its eventual eradication. Why have we not eliminated other pathogens in a similar manner?

8. Rhogam is an example of passive immunization. Injections can prevent the birth of "blue babies" having hemolytic disease of the newborn. Obtain literature about the use of rhogam from the public health department and report on the mechanism used to prevent Rh sensitization.

Challenge Questions

1. If a previously unknown pathogen has caused several recent deaths in a small mountain town, what measures would you take to contain the outbreak and determine its source?

2. If the above outbreak was found to be caused by a tick-borne bacteria, what additional precautions would you take?

3. Using your knowledge of the immune response, what is the main reason why organ transplants fail?

4. Under what conditions are monocytes converted to macrophages?

5. How do memory cells react to pathogens they are designed to detect?

6. What kind of immune responses would you expect from a blood transfusion?

LESSON
19

Animals: Respiration

Assignments

For the most effective study of this lesson, we suggest that you complete the assignments in the sequence listed below:

Before Viewing the Video Program
- Read the Overview and Learning Objectives for this lesson. Use the Learning Objectives to guide your reading, viewing, and thinking.

- Read Chapter 35, "Respiration," pages 582–597, in the Starr textbook.

  Or read, in the Starr/Taggart textbook, Chapter 41, "Respiration," pages 690–709.

- Read the Viewing Notes in this lesson.

View the "Respiration" Video Program
After Viewing the Video Program
- Briefly note your answers to the questions at the end of the Viewing Notes.

- Review all reading assignments for this lesson, especially the Chapter 35 Summary on page 596 of the Starr textbook (Chapter 41 Summary on page 708 in the Starr/Taggart textbook) and the Viewing Notes in this lesson.

- Write brief answers to all the Review Questions at the end of Chapter 35 in the Starr textbook (Chapter 41 in the Starr/Taggart textbook) to be certain you understand the text material.

- Complete the Review Activities in this guide to reinforce your understanding of important terms and concepts. Check your answers with the Answer Key and review when necessary.

- Take the Self-Test in this guide to measure your achievement of the Learning Objectives. Check your answers with the Answer Key and review when necessary.

- Complete the Using What You've Learned activities and any other activities and projects assigned by your instructor.

Overview

You began your study of animal structure and function by looking at some basic supporting tissues and how movement is produced. Your focus then shifted to the circulatory system and the mechanisms of immunity. In Lesson 19, you will explore the concepts and implications of animal respiration.

The lesson begins by emphasizing the importance of respiration in the production of energy necessary for basic animal functions. Essentially, a system of gas exchange allows oxygen to be delivered to the cells where the energy is produced. The same system also provides a means for a waste product, carbon dioxide, to be removed from the interstitial fluid and later eliminated. In your study of respiration, you will be introduced to the factors that affect efficient respiration and to the concept of pressure gradients, which are necessary for the system to work.

Interesting adaptations related to gas exchange can be seen in the invertebrates ranging from simple diffusion to a complex series of tubes used for tracheal respiration in insects. Next, gills and lungs are examined as we determine how vertebrates such as fishes and mammals have solved their need for more oxygen in a variety of environments. The human respiratory system is then presented to demonstrate how the physical structures support the cycle of gas exchange and transport. You'll learn about the nature of the respiratory surface where the exchange of gases occurs, and about the role of the circulatory system in gas transport. You'll learn about the role of the nervous system in monitoring and regulating breathing.

As the lesson continues, a clearer picture of the complexity of gas exchange emerges. The structure and function of red blood cells, circulatory gradients, and the effects of other gases such as nitrogen are discussed. Finally, the effects of disease and smoking are shown as examples of the fragility of the respiratory system and our own role in preventing serious health problems.

Learning Objectives

When you have completed all assignments for this lesson, you should be able to:

1. Identify the gases that most organisms exchange with their environment, discuss the different factors which affect the rate of this exchange, and describe the typical respiratory surface where the exchange occurs.

2. Compare gas exchange in unicellular and multicellular organisms, noting the challenges faced by larger body sizes.

3. Describe the different types of gas exchange structures, and match each of them with an animal that uses it.

4. Describe the structure of the human respiratory system, and explain how breathing cycles air to and from respiratory surfaces.

5. Explain how the respiratory system works with the circulatory system to transport gases to and from body tissues for cellular respiration.

6. Describe how different blood compounds assist gas transport, noting the role of hemoglobin and the bicarbonate reaction.

7. Discuss how the body's nervous system monitors and regulates breathing.

8. Describe some common respiratory disorders, noting how smoke and other airborne pollutants can affect respiratory tissue.

9. Describe some of the methods and tools physiologists use to study respiratory systems.

Viewing Notes

The video program for Lesson 19 graphically illustrates the need for an efficient mode of gas exchange throughout the animal kingdom. As you will see, the diversity of adaptations found in nature is directly related to the metabolic needs and energy demands of each organism. Dr. Karen Martin will tell you about the oxygen and carbon dioxide cycle and how simple diffusion can be enhanced by various structures and methods of gas transport. Watch as she uses the case of the sculpin fish to present an example of an oxygen exchange system.

In the next part, "Lung Power," we shift gears by joining athlete D. J. Peterson as he trains for a triathlon—a difficult combination of running, biking and swimming. The effects of these exercises on the respiratory system are explained by Dr. Christopher Cooper who also provides details about how the gases are exchanged within the human body. Computer animation showing oxygen transport at the cellular level helps to clarify these concepts. Dr. Cooper shows how careful measurements of gases and the products of metabolism during exercise give clues to the optimum efficiency of the body systems as they work together.

During the final part, entitled "Hold Your Breath," Dr. Gerald Kooyman investigates the ability of Weddell seals and Emperor penguins to take long underwater dives. You will discover the unique ways in which these and other animals have pushed the limits using "adaptive respiration." You will also gain an appreciation for some of the methods scientists use to study the process of respiration and the various forms it takes.

As you watch the video program, consider the following questions:

1. If gases are actively transported to the tissues, why is diffusion still necessary in the exchange process?
2. What are some factors which limit the use of gills to obtain oxygen?
3. Would you expect tidepool animals to be generally more active or less active at night, considering their respiratory needs?
4. Knowing that skin can be used as a respiratory surface, why is only a small portion of the skin actually used for this purpose?
5. What main role does the diaphragm play in breathing?
6. What are the implications of exceeding the "anaerobic" threshold?
7. How does measuring the metabolic rate relate to the length of time a seal or penguin can stay underwater?
8. According to Dr. Kooyman, what are the three compartments where oxygen is stored?
9. Could human beings ever "breathe" underwater or in space?
10. How do you think air pollution or smoking has changed the way animals have adapted to their respiratory environment?

Review Activities

Matching

Match the respiratory system pictured below with the terms that follow. Check your answers with the Answer Key and review any terms you missed.

I.

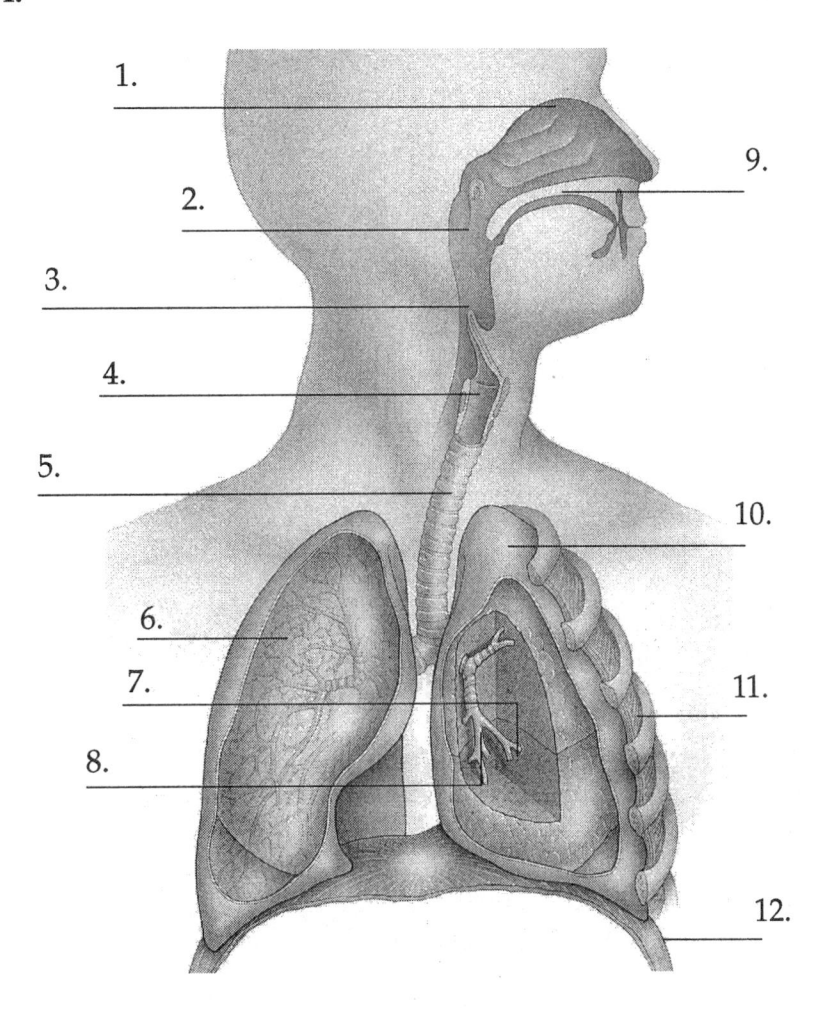

a. alveoli
b. bronchial tree
c. diaphragm
d. epiglottis
e. intercostal muscles
f. larynx
g. lung
h. nasal cavity
i. oral cavity
j. pharynx
k. pleural membranes
l. trachea

Lesson 19 / Animals: Respiration   239

Match the terms listed below with the definitions that follow. Check your answers with the Answer Key and review any terms you missed.

II.

____ 1. tracheal respiration
____ 2. oxyhemoglobin
____ 3. countercurrent flow
____ 4. integumentary exchange
____ 5. hemoglobin
____ 6. gill
____ 7. respiration
____ 8. myoglobin
____ 9. lung
____ 10. bronchus

a. found normally in muscle tissue, this molecule may store a significant amount of oxygen in many species including aquatic mammals

b. respiration accomplished by means of oxygen and carbon dioxide diffusion across a moist, thin layer of vascularized epidermis

c. a method used to enhance oxygen extraction by taking advantage of gradients found in two fluids moving in opposite directions

d. an organ found in both invertebrates and vertebrates where gas exchange takes place on a thin, moist layer of vascularized epidermis

e. a molecule of hemoglobin which has bound to oxygen after reaching the lungs

f. respiration accomplished by means of tubes carrying gases from the integument to the tissues in spiders, insects, and other animals

g. a protein molecule containing oxygen-binding sites and iron usually found within the red blood cell

h. the process of providing oxygen to metabolically active cells and removing carbon dioxide from the area

i. an organ found in birds, mammals, amphibians, and several fish containing an internal cavity or sac of respiratory surfaces for the purpose of gas exchange

j. a branching of the trachea leading to the lungs of most vertebrates

III.

____ 1. carbonic anhydrase   ____ 6. transport pigment
____ 2. diffusion   ____ 7. carbon monoxide
____ 3. oxygen   ____ 8. partial pressure
____ 4. lactic acid   ____ 9. bicarbonate
____ 5. carbon dioxide   ____ 10. nitrogen

a. molecule capable of binding gases used in respiration as well as giving blood cells their characteristic color such as red or blue

b. enzyme contained within red blood cells capable of converting carbon dioxide to carbonic acid which then dissociates to bicarbonate

c. a primary waste product of cellular metabolism which can be removed by various mechanisms of gas exchange and chemical conversions

d. process of active or passive molecular movement along specific concentration gradients

e. a primary waste product of anaerobic cellular metabolism which can accumulate and reduce the efficiency of further energy production

f. molecule resulting from carbon dioxide conversion and dissociation with the net effect of maintaining the concentration gradients necessary for efficient respiration

g. necessary for efficient aerobic cellular metabolism and is transported to the tissues by various mechanisms including the ability to bind to the hemoglobin molecule

h. molecule that binds to the hemoglobin molecule with an affinity 200 times that of oxygen, resulting in potential oxygen deprivation in the tissues

i. the most abundant gas in Earth's atmosphere which, if too highly concentrated in the human body, can cause a condition known as the "bends"

j. contribution of a gas to the total concentration gradient across a respiratory membrane

Completion

Fill in each blank with the most appropriate term from the list for that paragraph. A term may be used once, more than once, or not at all. If a question requires two or more answers in succession, they may be in any order (unless the question indicates otherwise). Check your answers with the Answer Key and review when necessary.

1. Respiration in the animal kingdom centers around obtaining _____ and energy production and removing _____ produced in the process. Essentially this involves gas _____ at the membrane level. The efficiency of this mechanism can be affected by the _____ ratio of an animal, limiting the _____ rate of the gases. Another factor enhancing the cycle is _____ which uses specialized adaptations such as _____ and lungs. There are also _____, such as hemoglobin in the _____ blood cell that _____ oxygen at one site and releases the gas in the _____. The _____ system works closely with the respiratory system in many animals and is important for the transport of gases and matching air _____ to _____ flow.

   red   bind
   white   exchange
   carbon monoxide   liquid to gas
   carbon dioxide   surface to volume
   gills   flow
   pressure   circulatory
   oxygen   digestive
   diffusion   ventilation
   blood   transport pigments
   tissues

2. Respiratory adaptations are diverse but attempt to do the same thing: _____ and transport oxygen to the tissues, provide for _____ exchange, and _____ of carbon dioxide. Many small invertebrates use diffusion through their _____ while aquatic species may use their _____ or organs called _____. In drier habitats, insects have developed a network of _____ while most _____ have small folded book _____. The vertebrates have also found unique solutions according to their _____ requirements and oxygen _____. Birds have _____ using a continuous flow across _____ surfaces. _____ such as salamanders use the skin to exchange gases or supplement less-efficient lungs. A _____ blood flow provides for a high rate of oxygen extraction in the gills of many _____. The mammals, including humans, have well-developed lungs with _____ for gas exchange. Thus, animals at all levels of complexity have been able to solve their respiratory needs.

spiders	skin
amphibians	air sacs
fish	gills
release	flatworms
lungs	integument
lateral	excretory
shells	hollow bones
alveoli	respiratory
obtain	gas
dispose	tubes
environment	countercurrent
energy	

Lesson 19 / Animals: Respiration

3. The human respiratory system is subject to many stresses and disease states. _____ and _____ pose real dangers to respiratory functions and are contributing factors in many other diseases. Other causative agents, such as allergens, viruses, and _____, can lead to specific problems. Initially, irritation of the lining of the _____ can cause _____ to develop. Chronic cases may progress into serious infections, lung damage, and even _____. A variety of substances are also inhaled such as _____, cocaine and opium derivatives, model glue and _____ distillates, airborne _____, and _____. Some may only cause a shortness of _____ while others may lead to asthmatic conditions and _____.

breath	petroleum
cigarette smoking	spicules
marijuana	fungi
dust	tuberculosis
bacteria	emphysema
fires	bronchioles
environmental pollution	bronchitis
stature	plastic
cancer	chlorine gas

Self-Test

Select the one best answer for each question. Check your answers with the Answer Key and review when necessary.

1. Fick's law indicates that the diffusion rate will increase across a respiratory surface when

   a. surface area is greater and the partial pressure gradient is smaller.

   b. partial pressure gradient is larger and surface area is greater.

   c. partial pressure gradient is smaller and surface area is decreased.

   d. surface area is decreased and the partial pressure gradient is larger.

2. Why are there so many different adaptations in the animal kingdom to provide sufficient gas exchange?

   a. variations in atmospheric composition

   b. variations in energy needs

   c. variations in molecular structure

   d. variations in the environment containing respiratory gases

3. Single-cell organisms and very thin animals can use diffusion to provide for gas exchange. What limiting factors prevent larger, multicellular animals from using this same mode of respiration?

   a. the surface-to-volume ratio is increased

   b. lack of efficient oxygen delivery to cells leads to insufficient energy production

   c. pressure gradients are too large between each surface

   d. more viscous cellular fluids slow the diffusion process

4. The most efficient respiratory system with respect to gas exchange can be found in which class of animals?

   a. reptiles

   b. mammals

   c. birds

   d. amphibians

5. When human beings inhale air into the lungs, which of the following occurs?

   a. volume in the chest cavity decreases

   b. pressure in the lungs is less than pressure in the thoracic cavity

   c. more carbon dioxide is released into the lymphatic system

   d. pressure in the lungs is lower than atmospheric pressure at the mouth

6. The circulatory system in humans helps to obtain oxygen for the tissues by
   a. providing lung capillaries in the alveoli for diffusion of oxygen from the air spaces into the red blood cells in close proximity.
   b. obtaining specific antibodies to bind oxygen.
   c. encasing the oxygen molecules inside small tubules.
   d. extracting hemoglobin from the alveolar lining.

7. The red blood cell performs all of the following functions EXCEPT
   a. transport of oxygen to the tissues using the circulatory system.
   b. enhanced conversion of carbon dioxide to bicarbonate using the carbonic anhydrase reaction.
   c. immune defense against bacteria in the alveoli.
   d. binding four molecules of oxygen for each molecule of hemoglobin inside the cell.

8. The nervous system can monitor and regulate the respiratory functions through
   a. detecting changes in blood pH.
   b. feedback of chemicals released in the lymph nodes.
   c. electrical activity in the pharynx and bronchioles.
   d. cranial nerves connected to the diaphragm.

9. Smoking and inhaling pollutants can affect tissues in the respiratory tract by all of the following EXCEPT
   a. damaging the cilia in the bronchioles.
   b. producing tumors initiated by carcinogens inhaled.
   c. lowering immune response to infection.
   d. increasing the elasticity of the alveolar sacs.

10. In order to better study respiratory systems, researchers might employ which of the following measurements?
    a. oxygen consumption
    b. pH changes in the blood
    c. lung volume capacity
    d. all of the above

Using What You've Learned

Based on your own interests or your instructor's requirements, complete one or more of the following activities.

1. Take a closer look at your home, garage, workplace or school environment. Make a list of all the airborne contaminants to which you are exposed. Are there ways to reduce or eliminate your exposure? If so, try implementing those solutions without breaking any laws!

2. Visit your favorite library and determine the main components in smog or other forms of air pollution. Compare this list with a list of known or suspected carcinogens. What can you conclude about air pollution?

3. Obtain the chemical structure of hemoglobin. Use diagrams to show the heme portion of the molecule separately, including ferrous iron and the four pyrrole rings that comprise protoporphyrin.

4. When you are at the market, ask your butcher if he has any samples of lung tissue to examine. Note the texture and density of the tissue. Why do you think the lung is structured this way and not simply like a balloon?

5. Design a new respiratory system that would be adaptive for both underwater habitats in the ocean as well as altitudes from sea level to 30,000 feet. What primary and secondary organs would be necessary? What kinds of physiological and chemical needs would be important to regulate? Which of the mechanisms you have studied so far might enhance your design?

Challenge Questions

1. How are the effects of the "bends" from deep-sea diving reversed or partially alleviated in a decompression chamber?

2. What does the term SCUBA stand for? What progress have we made since Jacques Cousteau invented this mechanical adaptation?

3. If you were dependent upon a respirator or "iron lung" to live, would you accept this restriction? If the answer is yes, under what circumstances and for how long would you continue using this form of respiration?

4. What are the normal reference ranges for arterial blood gases for humans? If a person were in keto-acidosis, how would those values change?

5. It has been noted that brain cells begin to die after being deprived of oxygen after only five minutes. How can you explain those cases where an individual has fallen to the bottom of an icy river or gone into respiratory arrest when covered by snow for 20 minutes or more and had no known permanent effects?

LESSON 20

Animals: Digestion and Fluid Balance

Assignments

For the most effective study of this lesson, we suggest that you complete the following assignments in the sequence listed below:

Before Viewing the Video Program

- Read the Overview and Learning Objectives for this lesson. Use the Learning Objectives to guide your reading, viewing, and thinking.

- Read Chapter 36, "Digestion and Human Nutrition," pages 598–617, and Chapter 37, "The Internal Environment," pages 618–629, in the Starr textbook.

   Or read, in the Starr/Taggart textbook, Chapter 42, "Digestion and Human Nutrition," pages 710–729, and Chapter 43, "The Internal Environment," pages 730–743.

- Read the Viewing Notes in this lesson.

View the "Digestion and Fluid Balance" Video Program

After Viewing the Video Program

- Briefly note your answers to the questions at the end of the Viewing Notes.

- Review all reading assignments for this lesson, especially the Chapter 36 and Chapter 37 Summaries on pages 616 and 628 in the Starr textbook (Chapter 42 and Chapter 43 Summaries on pages 728 and 742 respectively in the Starr/Taggart textbook) and the Viewing Notes in this lesson.

- Write brief answers to all the Review Questions at the end of Chapters 36 and 37 in the Starr textbook (Chapters 42 and 43 in the Starr/Taggart textbook) to be certain you understand the text material.

- Complete the Review Activities in this guide to reinforce your understanding of important terms and concepts. Check your answers with the Answer Key and review when necessary.

- Take the Self-Test in this guide to measure your achievement of the Learning Objectives. Check your answers with the Answer Key and review when necessary.

- Complete the Using What You've Learned activities and any other activities and projects assigned by your instructor.

Overview

In the last lesson, you learned about the exchange of oxygen and carbon dioxide that is the basis of respiration and fundamental to human survival. Yet oxygen alone is not sufficient to sustain human life. In order to fuel the energy-producing activities of the cell, water and other nutrients must be taken in from the external environment, often in forms that are chemically useless in their natural state.

To extract the nutrients they need and deliver them to the cells, animals and other complex organisms have developed sophisticated mechanisms to ingest and break down foods into metabolically useful constituents. This is the process of digestion, which culminates in the absorption of those nutrients needed most by the cells. While the digestive systems of higher life forms involve a variety of organs and tissues, they are essentially a series of food processors dedicated to the progressive mechanical and chemical decomposition of food.

While extremely efficient, the cells of our body are unable to use all of the foods and fluids we take in. Also, there are by-products of cellular metabolism that are useless, even dangerous to the organism. To deal with the buildup of these leftover waste materials, and preserve the normal volume and composition of their internal fluid environment, animals have evolved equally sophisticated systems of elimination. In human beings, this system includes the kidneys and vessels of the urinary tract.

The complementary processes of digestion and elimination are the primary subjects of Lesson 20. Through comparisons with other members of the animal kingdom and in-depth exploration of the anatomy and physiology of these systems in human beings, you will come to understand and appreciate the elaborate strategies nature has invented to regulate the internal environment, preserve homeostasis, and ensure survival.

Learning Objectives

When you have completed all assignments for this lesson, you should be able to:

1. Define *ingestion*, *digestion* (physical and chemical), *absorption*, and *egestion*. Describe where each process occurs in single-celled organisms and in organisms with *incomplete* and *complete* digestive systems.

2. Describe the structure of the human digestive system and specify the function of each of its regions, giving special attention to the role of *villi*.

3. Identify each of the major digestive secretions, noting the enzymes and other chemicals of significance and their roles in digestion.

4. Explain the adaptive advantages of the specialized digestive organs of birds and ruminants.

5. Discuss the concept of the *food pyramid*, and explain how it would help an individual meet nutritional requirements for carbohydrates, proteins, fats, vitamins, and minerals.

6. Explain in general terms how the chemical composition of *extracellular fluid* is maintained by mammals, emphasizing the different routes for gains and losses, and identifying the most common metabolic wastes emitted.

7. Describe the gross anatomy of the *urinary system* and explain how the various components work together to filter, reabsorb, and secrete water and solutes to form urine.

8. Identify the location and parts of a *nephron*.

9. Discuss the kidney's role in maintaining homeostasis using the examples of hormone-induced regulation of water and sodium levels.

10. Compare the human urinary system with that of other animals, noting the different composition of wastes excreted and how they are formed.

11. Describe some of the methods scientists use to study digestive and urinary systems.

Viewing Notes

Program 20, "Digestion and Fluid Balance," provides numerous examples to illustrate how animals regulate (control) food and fluid input/output to maintain the internal homeostasis. From hummingbirds to human beings, each species has evolved multiple mechanisms to take in, process, and absorb essential fluids and nutrients. And they have evolved equally sophisticated mechanisms to consolidate and get rid of unwanted waste material.

Take the hummingbird, for example. To support its high metabolic rate, the hummingbird must consume and quickly process large amounts of nectar. Just how the hummingbird does this is a subject of study by Professor Carol Beuchat. By joining Professor Beuchat in her fieldwork, you will begin to understand how the complementary processes of digestion and elimination work and appreciate the methods scientists use to study these processes.

In Part Two, "Doggie Bag, Please," the importance of nutrition and nutritional requirements is discussed through interviews with pet owners and Kal Kan Foods Manager, Dr. J. H. Sokolowski. This discussion will prepare you for an in-depth tour of the digestive system, where you will see how nutrients are ultimately broken down and absorbed by the organs of the gastrointestinal tract. Your tour guide is Austin K. Mircheff, Professor of Physiology and Biophysics. Using computer animation and familiar analogies, he will walk you through each step of the digestive process, pointing out key fluids and enzymes along the way.

A similar tour of the kidneys and urinary system awaits you in Part Three, "When the System Fails." Drawing on the case of a woman suffering from kidney failure, Professors Alicia McDonough and Cyril Barton will show you how the elaborate filtration system in our kidneys works and what happens when that system become inoperative. In the process, you'll learn about the various hormonal and nutritional factors that affect kidney function and, ultimately, the stability of the internal fluid environment.

As you watch the video program, consider the following questions:

1. What kinds of mechanisms have animals evolved to take in food and water? What kinds of mechanisms have they evolved to eliminate solid and liquid waste materials?
2. What forms do these waste materials take?
3. How do the functions of the digestive system and urinary system complement each other?
4. What are the key organs and fluids of the digestive system? What are the various functions of these organs and fluids?
5. What kinds of nutrients are essential to the health and well-being of human beings? Are they the same for other animals?
6. What are the key components of the urinary system? What are the various functions of these components?
7. How do hormones help regulate the concentration of blood and interstitial fluid?

Review Activities

Matching

Match the cell structures pictured below with the terms that follow. Some terms may be used more than once. Check your answers with the Answer Key and review any terms you missed.

I.

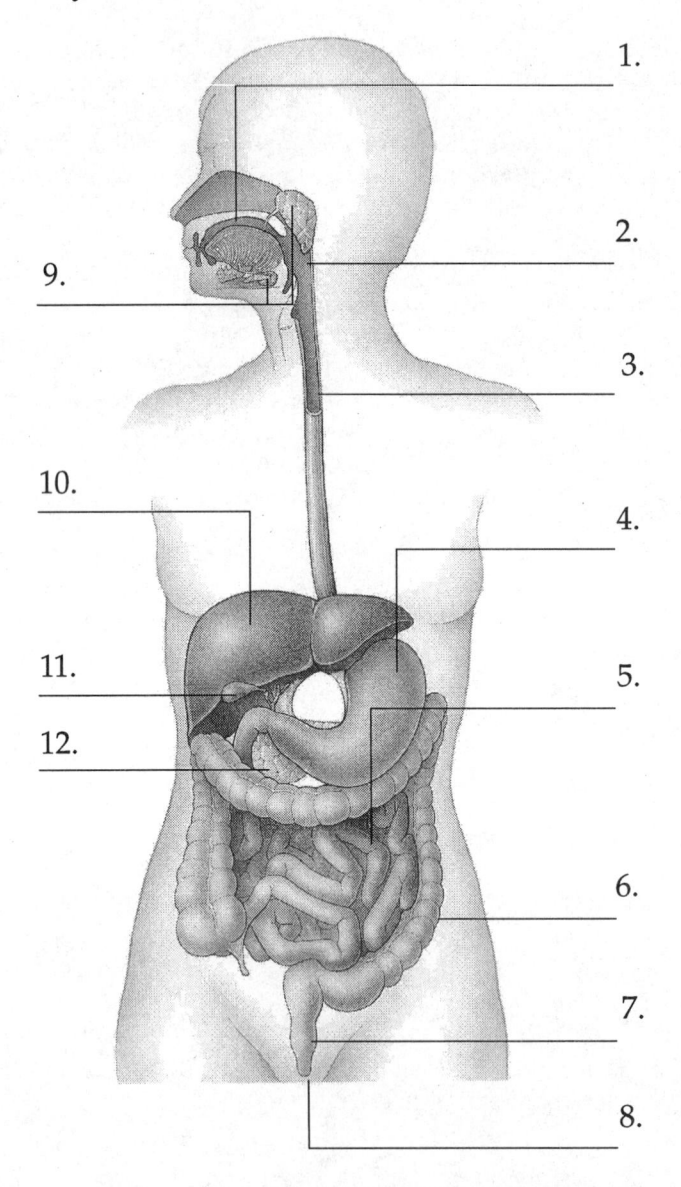

1.
2.
3.
4.
5.
6.
7.
8.
9.
10.
11.
12.

- a. anus
- b. colon
- c. esophagus
- d. gallbladder
- e. liver
- f. oral cavity
- g. pancreas
- h. pharynx
- i. rectum
- j. salivary glands
- k. small intestine
- l. stomach

252   Lesson 20 / Animals: Digestion and Fluid Balance

II.

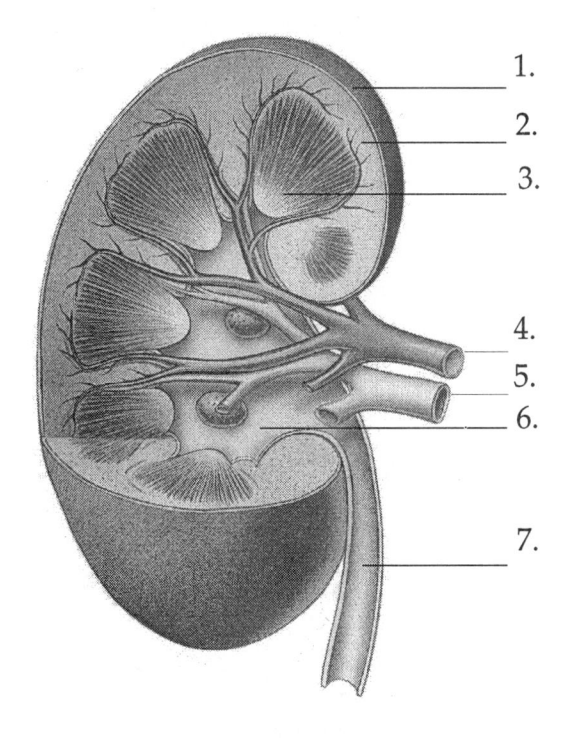

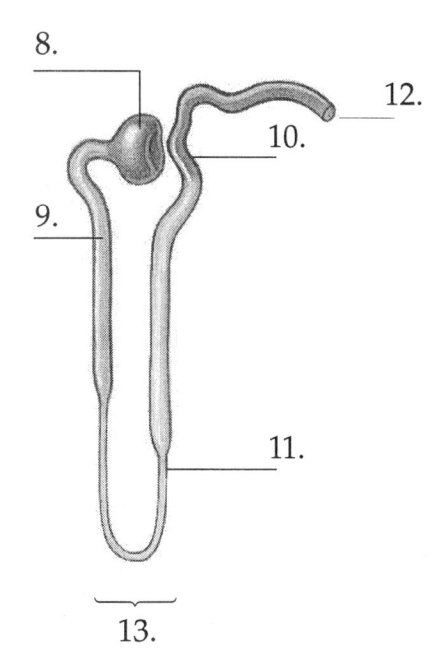

1.
2.
3.
4.
5.
6.
7.
8.
9.
10.
11.
12.
13.

a. collecting tubule
b. distal tubule
c. glomerulus
d. kidney cortex
e. kidney medulla
f. loop of Henle
g. nephron
h. proximal tubule
i. renal artery
j. renal capsule
k. renal pelvis
l. renal vein
m. ureter

Now match the terms listed below with the definitions that follow. Check your answers with the Answer Key and review any terms you missed.

III.

____ 1. antidiuretic hormone   ____ 6. extracellular fluid
____ 2. amino acid   ____ 7. fatty acid
____ 3. aldosterone   ____ 8. gastric fluid
____ 4. bile   ____ 9. saliva
____ 5. chyme   ____ 10. urine

a. alkaline solution secreted by the liver that aids in the digestion and absorption of fats

b. hormone secreted by the adrenal cortex that helps regulate sodium absorption

c. the primary organic compound of polypeptide chains that make up protein-based nutrients

d. alkaline secretion of water, mucin, protein, and salt that lubricates ingested food and begins the breakdown of starches

e. fluid formed by filtration, reabsorption, and secretion by the kidneys

f. hormone produced by the hypothalamus that stimulates reabsorption in the kidneys

g. semifluid mass of partly digested food expelled by the stomach into the duodenum

h. monocarboxylic acids that occur naturally in the form of glyceride-based nutrients

i. fluid that encompasses the plasma of the blood and interstitial fluid that occupies the space between the cells and tissues

j. thin, watery digestive fluid secreted by glands in the mucous membrane of the stomach

IV.

___ 1. conduction ___ 6. excretion
___ 2. convection ___ 7. filtration
___ 3. digestion ___ 8. ingestion
___ 4. emulsification ___ 9. reabsorption
___ 5. evaporation ___ 10. secretion

a. a suspension of fat droplets coated with bile salts

b. the transfer of water and usable solutes out of nephrons and into capillaries

c. the expulsion of waste material

d. the process of breaking down food into simpler chemical compounds for absorption

e. the transfer of heat through fluid circulation

f. the process of taking in food and water for digestion

g. the release of fluids and solutes into the lumen of an organ or vessel

h. the transfer of heat by means of a conducting material

i. the process by which blood pressure forces water and solutes out of a kidney's glomerular capillaries and into the Bowman's capsule

j. conversion of a substance from the liquid to the gaseous state

Completion

Fill in each blank with the most appropriate term from the list for that paragraph. A term may be used once, more than once, or not at all. If a question requires two or more answers in succession, they may be in any order (unless the question indicates otherwise). Check your answers with the Answer Key and review when necessary.

1. The preservation of _____ depends in part on an organism's ability to regulate the chemical composition of its _____. It does this by ingesting food and fluids, retaining the _____ it needs, and eliminating _____. In humans, nutritional requirements include daily portions of complex _____, _____, _____, and certain _____ and _____. These _____ are extracted from digested food and absorbed into the body through the _____ of the _____. While most nutrients enter the _____ for transport throughout the body, _____ are transported by the _____ for storage in _____. By-products of the digestive process in the form of _____, _____, and _____ are excreted, while excess _____ can be eliminated through _____, _____ (sweat), or _____.

adipose tissue
ammonia
bloodstream
carbohydrates
colon
evaporation
excretion
fats
extracellular fluid
homeostasis
internal environment
liver
lymph system

minerals
nutrients
proteins
respiration
small intestine
transpiration
villi
vitamins
urea
uric acid
urinary system
waste materials
water

256   Lesson 20 / Animals: Digestion and Fluid Balance

2. _____ provide a mechanism for the intake, transport, processing, and storage of food and water. In humans, the mechanical and chemical breakdown of food begins in the _____, where the enzymes of _____ begin the work of digesting _____. _____ also lubricates the food for its journey down the _____. When it reaches the _____, _____ mixes with food and kills many pathogens. _____ digestion also begins here. As the food enters the _____, it is mixed with digestive secretions from the _____, _____, and _____. By the time the food nears the small intestine, most nutrients have been _____ and _____. The remaining undigested matter is concentrated in the _____ and eventually _____ through the _____.

absorbed	liver
alimentary canal(s)	oral cavity
bile	pancreas
colon	pharynx
chyme	polysaccharide(s)
digested	protein(s)
digestive system(s)	rectum
esophagus	saliva
excreted	secreted
gallbladder	small intestine
gastric fluid	stomach
gastrointestinal tract(s)	triglyceride(s)
ingested	

Lesson 20 / Animals: Digestion and Fluid Balance

3. Maintaining balance in the volume and composition of the _____ is the primary role of the _____. In humans, the system consists of _____ which filter water, minerals, organic wastes and other substances out of the blood. The functional units of the kidney, the _____, process these materials in three stages. In the first stage, blood enters the _____ where _____ and _____ are _____ into the _____. In stage two, most of the water and solutes are reabsorbed into _____ adjacent to proximal and distal _____. In the third stage, water and solutes not reclaimed by the capillaries are _____ into the _____ where they are ultimately expelled as _____ through the _____, _____, and _____.

blood
blood cells
Bowman's capsule
capillaries
extracellular fluid
filtered
glomerulus
intracellular fluid
kidney(s)
loop of Henle
nephron
proteins

reabsorption
renal capsule
secreted
small solutes
tubules
ureter
urethra
urinary bladder
urinary system
urinary tract
urine
water

Self-Test

Select the one best answer for each question. Check your answers with the Answer Key and review when necessary.

1. The primary difference between the incomplete digestive system of invertebrates such as flatworms and the complete digestive system of more complex vertebrates is that

    a. in flatworm digestive systems, all waste materials are recycled; in vertebrate digestive systems, they must be excreted.

    b. the digestive system of complex vertebrates has two openings, whereas the digestive system of flatworms has only one.

    c. the digestive system of flatworms is designed to process simple carbohydrates, not complex carbohydrates as in vertebrates.

    d. the digestive system of flatworms processes food, fluids, and oxygen; in vertebrates, the processing of food and oxygen is handled by separate systems.

2. Villi play an important role in digestion because they

    a. secrete the digestive fluids that break down food into nutrients.

    b. supply blood to all the organs of the gastrointestinal tract.

    c. move undigested food into the colon for storage.

    d. increase the surface area available for nutrient absorption.

3. Of the enzymes listed below, the digestive enzyme that is important in carbohydrate digestion is

    a. lipase.

    b. pepsin.

    c. trypsin.

    d. pancreatic amylase.

4. Ruminants such as cows and antelopes have developed more than one stomach in order to

    a. prolong the mechanical and chemical breakdown of plant material.

    b. extract different nutrients from cellulose.

    c. survive for longer periods between feedings.

    d. replace the functions of the small and large intestines.

5. Added fats and simple sugars are at the top of the food pyramid because
    a. they are the most important foods in the human diet.
    b. they are the most difficult foods to digest.
    c. they should be consumed only in limited quantities.
    d. they are needed to digest proteins and complex carbohydrates.

6. The volume and composition of the internal fluid environment are maintained through all of the following EXCEPT
    a. the evaporation of sweat.
    b. the ingestion of solid food.
    c. the inhalation of water vapor.
    d. the excretion of urine.

7. In the urinary system, the formation of urine occurs in the
    a. kidney.
    b. ureter.
    c. urethra.
    d. urinary bladder.

8. Within the nephron, reabsorption of water and solutes occurs in the
    a. proximal and distal tubules.
    b. Bowman's capsule.
    c. glomerulus.
    d. renal pelvis.

9. Aldosterone aids in the regulation of sodium absorption by
    a. reclaiming sodium ions awaiting excretion in the urinary bladder.
    b. decreasing the permeability of the collecting ducts to water.
    c. chemically bonding with sodium ions and actively transporting them.
    d. stimulating increased reabsorption of sodium in the distal tubules.

10. The loops of Henle in a kangaroo rat's kidneys are especially long. This enables the kangaroo rat to
    a. lower its osmotic gradient relative to the external environment.
    b. dilute its concentrated urine before it damages surrounding tissues.
    c. dissipate heat and reduce its internal body temperature.
    d. maximize the reabsorption and retention of water.

Using What You've Learned

Based on your own interests or your instructor's requirements, complete one or more of the following activities.

1. Using a flexible inner tube and a ball of appropriate size, demonstrate how food is moved through the digestive tract through peristalsis.

2. Using a model or illustration, trace the progression of a mouthful of food as it passes through the entire gastrointestinal tract. Explain the role of key organs, fluids, and enzymes in the mechanical and chemical breakdown of food, the transport of nutrients, and the formation and elimination of waste materials.

3. Identify an eating disorder from your own experience or that of a family member or friend. Visit the library and scan recent science publications for current research on the disorder. Select the most interesting article and construct a brief report summarizing the pathophysiology, principal symptoms, clinical course, and recommended treatment for the disorder.

4. Identify a particular diet plan you have tried or one that has been used by a family member or friend. Again, scan recent research publications for a scientific analysis of the diet and its effectiveness. Select the most interesting article and construct a brief report summarizing the physiological basis for the diet, its advantages and limitations, and any safety concerns.

5. Using the analogy of a common drip-type coffeemaker, describe the function of nephrons. If possible, show how each part of the nephron compares with the corresponding part of the coffeemaker.

6. Urinalysis is an important part of a complete health assessment. Visit a medical laboratory and ask a clinical lab technician to explain what is included in such an assessment and what the results mean. Share your findings in a written or an oral presentation.

7. Speak to a physician or clinical specialist about the function and operation of a dialysis machine. What aspects of normal urinary function does the machine replace or supplement? Share your findings in a written or an oral presentation.

8. Read the contents labels on food products you consume and compare your intake of sugar, fat, and protein with recommended daily requirements.

Challenge Questions

1. How do over-the-counter antacids and other medicines reduce the discomfort of heartburn and other digestive ailments?

2. Why does drinking coffee or tea often increase the sensation of hunger?

3. Why are dietary fibers such as bran important in the human diet?

4. What kinds of roles do vitamins and minerals play in human digestion? Which ones are essential to health?

5. While considered bizarre by Western cultures, the practice of consuming one's own urine is practiced in Eastern cultures such as India. What are some possible benefits and risks of this practice?

6. Medical science has shown that humans can survive with one kidney. Why do we have two?

7. What are kidney stones, and why do they form?

8. Why is it considered unhealthy to drink seawater?

9. Eating salty foods usually results in temporary weight gains. Why?

LESSON 21

Animals: The Neural Connection

Assignments

For the most effective study of this lesson, we suggest that you complete the assignments in the sequence listed below:

Before Viewing the Video Program

- Read the Overview and Learning Objectives for this lesson. Use the Learning Objectives to guide your reading, viewing, and thinking.

- Read Chapter 29, "Integration and Control: Nervous Systems," pages 472–493, and Chapter 30, "Sensory Reception," pages 494–505, in the Starr textbook.

    Or read, in the Starr/Taggart textbook, Chapter 34, "Information Flow and the Neuron," pages 558–569; Chapter 35, "Integration and Control: Nervous Systems," pages 570–587; and Chapter 36, "Sensory Reception," pages 588–607.

- Read the Viewing Notes in this lesson.

View "The Neural Connection" Video Program
After Viewing the Video Program

- Briefly note your answers to the questions at the end of the Viewing Notes.

- Review all reading assignments for this lesson, especially the Chapter 29 and Chapter 30 Summaries on pages 492 and 504 in the Starr textbook (Chapter 34, Chapter 35, and Chapter 36 Summaries on pages 568–569, 586, and 606 respectively in the Starr/Taggart textbook) and the Viewing Notes in this lesson.

- Write brief answers to all the Review Questions at the end of Chapters 29 and 30 in the Starr textbook (Chapters 34, 35, and 36 in the Starr/Taggart textbook) to be certain you understand the text material.

- Complete the Review Activities in this guide to reinforce your understanding of important terms and concepts. Check your answers with the Answer Key and review when necessary.

- Take the Self-Test in this guide to measure your achievement of the Learning Objectives. Check your answers with the Answer Key and review when necessary.

- Complete the Using What You've Learned activities and any other activities and projects assigned by your instructor.

Overview

In previous lessons, you've learned about the various systems that have evolved in animals to manage the exchange of oxygen and carbon dioxide, the processing and distribution of nutrients, and the elimination of waste materials. While these life-sustaining functions continue, for the most part, without our conscious awareness or control, their activity is in fact carefully coordinated and monitored by perhaps the most sophisticated system of all—the nervous system.

Lesson 21 will introduce you to the sophisticated array of sensory receptors, nerves, and control centers called the nervous system. You'll learn about the various mechanisms that have evolved in the animal kingdom to detect changes in the environment—whether they be chemical changes, temperature changes, or changes in quality of light energy. You'll see how the minutest changes trigger the chemical reactions required to initiate nerve impulse transmission—and how the chain reaction of electrical impulses travels from nerve to nerve until it reaches the spinal cord or beyond. You'll learn about the role of the spinal cord in integrating and responding to incoming messages through established reflex pathways and the role of the brain in providing higher-level processing and control functions.

Your introduction to the neural connection will include some history on how the vertebrate nervous system evolved and how it varies in form and function among the species we know today. It will also include a brief survey of various drugs and their effects on the nervous system. You will find that study of these effects provides important insights into the normal function of the human nervous system.

Learning Objectives

When you have completed all assignments for this lesson, you should be able to:

1. Describe the general purpose and structure of *neurons*, and state their relationship to a *nerve*.

2. Explain how a neuron "at rest" maintains a voltage difference across its membrane and how this changes during an *action potential* (nerve impulse). Note how conduction is modified with the presence of *myelin*.

3. Describe the general structure of a *synapse*, and explain how it transmits information.

4. Identify the components and operation of a *reflex pathway*, including receptor, sensory neuron, interneuron, motor neuron and effector (muscle or gland). Using an example reflex, explain how sensory inputs are relayed to the *central nervous system*.

5. Describe the organization and structure of the vertebrate nervous system, and contrast the basic roles of its major divisions and subdivisions.

6. Identify the different types of *sensory receptors* found in the animal kingdom, and offer examples of each.

7. Describe how the brain and sense organs interact to interpret sensations and distinguish their relative intensity.

8. Summarize how the human ear captures sound waves and translates them into signals the brain can interpret.

9. Describe the structure of the vertebrate eye, and explain how it focuses and translates light into signals the brain can interpret.

10. Discuss the methods and techniques scientists use to study the nervous system and to determine how it functions.

Viewing Notes

To supplement your study of the nervous system, the video portion of this lesson will provide a rich source of expertise and examples from the world of biology. You will hear from Dr. James McGaugh, Director of the Center for the Neurobiology of Learning and Memory, as he describes the common mechanisms that characterize the detection and response to environmental stimuli throughout the animal kingdom. You'll hear from Dr. Peter Narins as he speaks of the supersensitive sensory receptors of the snake and spiny lobster and the receptors that all animals use to sense pain.

With these examples as a foundation, you will be transported to the cellular level to see how the transmission of nerve impulses actually occurs. You'll watch the movement of ions in and out of the neuron that produces the depolarization necessary to trigger nerve impulse transmission. With Dr. Narins to guide you, you'll begin to appreciate how this biochemical scenario plays out a million times a day to produce the internal and external sensations and reactions we take for granted.

Part Two of the video program examines more closely the chemical mediators that facilitate the transmission of impulses across the synapses between nerve cells. Through the experiences of polo instructor Kathy Batchelor, you'll discover how drugs such as morphine operate on the synapses to block the transmission of pain. Part Three explores the integrative function of the nervous system, which occurs in the brain and spinal cord. Dr. James Hicks explains the difference between simple reflex pathways—which are routed through the spinal cord—and the more sophisticated information processing that occurs only in the brain.

You'll see reflex pathways dramatically demonstrated as police cadets complete a firearms practice course. Just as the somatic nervous system is trained to react quickly to external events, so the autonomic nervous system is designed to respond quickly to the body's internal changes. Both systems rely on the ability of the central nervous system to integrate billions of bits of information and provide appropriate responses. Through computer-generated imagery, you'll see which areas of the brain are responsible for these control functions and for conscious thought.

As you watch the video program, consider the following questions:

1. What kinds of mechanisms do animals use to detect change in their environment?

2. How do animals sense pain and pleasure?

3. What is the basic cellular unit of the nervous system, and what is its structure?

4. What is the chemical basis for the electrical discharge that characterizes a nerve impulse?

5. What is myelin, and how does it facilitate nerve impulse transmission in vertebrates?

6. How do neurotransmitters and neuromodulators affect nerve impulse transmission across the synapse?
7. How are sensory nerve impulses so quickly integrated and translated into action?
8. What is the difference between the central and peripheral nervous system? Between the somatic and autonomic nervous system?
9. How do reflexes work?
10. How is the brain structured?

Review Activities

Matching

Match the terms listed below with the definitions that follow. Check your answers with the Answer Key and review any terms you missed.

I.

___ 1. action potential
___ 2. axon
___ 3. dendrite
___ 4. ganglion
___ 5. nerve
___ 6. neuron
___ 7. neuroglial cell
___ 8. neurotransmitter
___ 9. sensory nerve
___ 10. somatic nerve
___ 11. synapse
___ 12. sympathetic nerve

a. the basic cell of the nervous system

b. a cluster of cell bodies of neurons in regions other than the brain and spinal cord

c. the long, cylindrical extension of a neuron that ends in the finely branched endings where neurotransmitters are released

d. a nerve cell that carries signals to move the head, trunk, and limbs

e. the gap between the output zone of one neuron and the input zone of an adjacent neuron

f. a cell that provides structural and metabolic support for neurons

g. a nerve cell that acts as a sensory receptor and relays signals to the brain

h. a nerve that generally acts to speed up the body during times of heightened awareness or excitement

i. abrupt reversal in the steady voltage difference across a neuron's plasma membrane

j. bundles of the axons of sensory and/or motor neurons that act as a communication line for the peripheral nervous system

k. any of the signaling substances secreted by neurons

l. the short, slender extension from the cell body of a neuron

II.

___ 1. central nervous system
___ 2. cerebellum
___ 3. cerebral cortex
___ 4. cerebrum
___ 5. hypothalamus
___ 6. limbic system
___ 7. medulla oblongata
___ 8. parasympathetic nerve
___ 9. peripheral nervous system
___ 10. pons
___ 11. reticular formation
___ 12. thalamus

  a. part of the forebrain that influences hunger, thirst, and sexual drive in response to visceral conditions

  b. part of the brainstem that contains the reflex centers for circulation, respiration, and other vital functions

  c. a nerve that carries signals to slow the body down and divert energy to basic life-sustaining activities

  d. part of the forebrain where the most complex information integration occurs

  e. the part of the brain that coordinates sensory input and relays signals to the cerebrum

  f. major network of interneurons in the brainstem that help control the activities of the nervous system

  g. the part of the brain that coordinates signals between the cerebellum and the brain's higher integrating centers

  h. surface layer of the brain hemispheres that interprets sensory input, integrates information, and coordinates motor responses

  i. the part of the brain that works in conjunction with the cerebral cortex to control emotions

  j. the system that includes the brain and spinal cord

  k. the system of nerves leading to and from the spinal cord and brain

  l. the part of the hindbrain that contains the reflex centers for maintaining posture and fine-tuning limb movements

III.

___ 1. chemoreceptor
___ 2. cone cell
___ 3. cochlea
___ 4. echolocation
___ 5. mechanoreceptor
___ 6. nociceptor
___ 7. photoreceptor
___ 8. rod cell
___ 9. retina
___ 10. thermoreceptor

a. sensory receptor that detects the chemical energy in ions or molecules dissolved in surrounding fluid

b. light-sensitive inner layer of the eye

c. receptor cell in the vertebrate eye that is sensitive to dim light

d. sensory receptor that detects any stimulus causing tissue damage

e. sensory receptor that detects energy associated with changes in pressure, position, or acceleration

f. sensory receptor that detects radiant energy associated with temperature changes

g. receptor cell in the vertebrate eye that is sensitive to intense light

h. sensory receptor that detects changes in light energy

i. a coiled, fluid-filled structure in the inner ear that converts changes in pressure into auditory signals

j. the use of ultrasound waves generated by an organism to establish relative position in the environment

Completion

Fill in each blank with the most appropriate term from the list for that paragraph. A term may be used once, more than once, or not at all. If a question requires two or more answers in succession, they may be in any order (unless the question indicates otherwise). Check your answers with the Answer Key and review when necessary.

1. The _____ of vertebrates is characterized by a bilateral network of sensory and motor _____ bundled together into _____. _____ nerves carry the signals for moving the head, trunk, and limbs, while _____ nerves carry signals for adjusting the function of vital organs. The autonomic nerves are either _____ (speed up body functions in times of high stimulation) or _____ (slow down body functions in times of low stimulation). The _____ consists of the brain and spinal cord. The _____ provides reflex pathways and integrates multiple signals for transmission to the _____. The _____ contains different structures for the coordination of sensory and motor input, control of metabolic and motor activities, and higher thought processes such as memory and abstract reasoning.

autonomic	neurons
brain	parasympathetic
central nervous system	peripheral nervous system
nerve net	somatic
nerves	spinal cord
sympathetic	synapse

2. Nerve signals are transmitted by _____ changes in the _____ between a neuron's input zone (_____ and _____) and output zone (_____). A disturbance at the plasma membrane causes _____ to enter the neuron. If the disturbance is great enough and enough ions cross over, the normal voltage difference across the membrane (_____) is abruptly reversed (_____). _____ that arrive at an output zone trigger the release of _____, which diffuse across the _____ and either stimulate or inhibit threshold in the adjacent cell. Between _____, _____ are used to restore and maintain the normal ion gradients.

action potential(s)	plasma membrane
axon endings	potassium ions
cell body	resting membrane potential
chemical	sodium ion(s)
dendrites	sodium-potassium pumps
ions	synapse
neurotransmitters	voltage

3. Vertebrates have developed a variety of specialized _____ to detect changes in light, heat, sound, and movement within the body as well as in the external environment. The intensity of the sensation is a function of _____ receptors are activated and _____ they fire. _____ sensations such as _____, _____, and _____ are detected by mechanoreceptors in the muscles, joints, and skin. Nociceptors alert the body to stimuli strong enough to cause _____, which is interpreted by the brain as _____. Thermoreceptors in the body's interior and skin surface detect _____ or _____. More complex organisms have evolved specialized acoustic receptors for _____ and photoreceptors for _____.

cold	position
hearing	pressure
heat	sensory neurons
how many	somatic
how often	tissue damage
motor	vision
movement	what kind
pain	when

Self-Test

Select the one best answer for each question. Check your answers with the Answer Key and review when necessary.

1. What is the basic unit of nerve impulse transmission in the human nervous system?
    a. ganglion
    b. axon
    c. neuron
    d. dendrite

2. Resting membrane potential changes to an action potential when
    a. the electrical polarity in and around a nerve cell abruptly reverses.
    b. sodium and potassium ions are suddenly released from a nerve cell.
    c. the chemical composition of the nerve's plasma membrane changes.
    d. the brain floods the bloodstream with neurotransmitters.

3. A neurotransmitter crossing the chemical synapse of a nerve cell will
    a. excite the adjacent cell to reach its action potential.
    b. inhibit the adjacent cell from reaching its action potential.
    c. do either a or b.
    d. have no effect on the adjacent cell.

4. In a reflex pathway, signals sent from a sensory receptor
    a. are transmitted to adjacent muscle cells, which contract.
    b. are sent to the spinal cord where they stimulate motor neurons.
    c. flood the area with neurotransmitters, which stimulate adjacent motor neurons.
    d. are sent to the brain for processing and response.

5. In the vertebrate nervous system, the component that contains the reflex centers for vital tasks such as respiration and circulation is the
    a. cerebellum.
    b. cerebral cortex.
    c. medulla oblongata.
    d. cerebrum.

6. In vertebrates, rods and cones are the primary sensory organs responsible for
    a. thermoreception.
    b. echolocation.
    c. mechanoreception.
    d. photoreception.

7. How do humans distinguish between sensations of different intensity?
    a. Intense stimuli trigger different receptors than weak stimuli.
    b. Action potentials vary in strength based on the intensity of the stimulus.
    c. Intense stimuli cause more sensory receptors to fire, and to fire more rapidly.
    d. Different neural pathways are stimulated by different types of stimuli.

8. Which of the following mechanisms explains how humans hear?
    a. Sound vibrations are translated into fluid pressure changes in the cochlea.
    b. Sounds of different pitches activate different types of audioreceptors in the ear.
    c. High-frequency sounds emitted from the ear bounce off objects.
    d. Hair cells in the semicircular canals are stimulated by sound waves.

9. Which of the following is a requirement for vertebrate vision?
    a. Different colors must be converted to white light.
    b. Light rays must be focused so they converge on the retina.
    c. The retina must be covered with the same kind of light receptors.
    d. The space within the eyeball must be free of any fluid material.

10. A common method that scientists use to learn about the function of the nervous system is to
    a. use chemical tracers to follow the transmission of nerve impulses.
    b. dissect and compare the brains of different species.
    c. study the effects of different drugs on behavior.
    d. construct robots that mimic key neurological functions.

Using What You've Learned

Based on your own interests or your instructor's requirements, complete one or more of the following activities.

1. Using modeling clay or another material, fashion a model of the human brain. Describe the role of each of the functional areas.

2. Using a reflex hammer or other stimulus, test the reflexes of another student or volunteer. If possible, time the reflex. Describe the probable circuit through which sensory impulses are translated into action.

3. Test the sensations of other students or volunteers by first blindfolding them, then checking their ability to locate the source of different sounds or smells. Have them hold their nose and test their ability to identify different foods. Propose mechanisms to explain how each of the senses tested performs its task. How do the senses work together to enhance each other's function?

4. Visit the library and gather information on a well-known neurological or neuromuscular disorder such as multiple sclerosis, cerebral palsy, or Parkinson's disease. Report on the causes of the disorder, the mechanisms that produce the classic clinical signs and symptoms, and current modes of treatment.

5. Using data gathered from the library, prepare an argument for legalizing or prohibiting the legal use of a particular drug. Explain the short-term and long-term effects of the drug, and evaluate the relative benefits and risks of using it.

6. Visit the library and research an alternative healing therapy, such as chiropractic, acupuncture, shiatsu, or reflexology. Describe the purpose of the therapy and by what physiological mechanisms it claims to achieve this purpose. How effective is the therapy? What are its limitations?

7. Ask your family physician to show you how an opthalmoscope and an otoscope work. Ask him to let you use these instruments to look into the eyes or ears of a volunteer. What do you see? How do these instruments help medical experts diagnose certain conditions?

Challenge Questions

1. How is that some people are tone deaf, while others have "an ear for music"? What is it about their nervous system that makes virtuosos so good at what they do?

2. Why does drinking large amounts of coffee and other caffeinated drinks make a person "jittery"? What is going on with the person's nervous system under these conditions?

3. Why is it that older people can remember the details of a distant event but are unable to recall what they had for dinner the night before?

4. What purpose do dreams serve? Do they reveal clues to the inner thoughts, concerns, or aspirations of the dreamer?

5. In humans, the cell bodies of neurons are not regenerated after injury or death. What adaptive advantages or disadvantages might this pose?

LESSON 22

Animals: Endocrine Control

Assignments

For the most effective study of this lesson, we suggest that you complete the assignments in the sequence listed below:

Before Viewing the Video Program
- Read the Overview and Learning Objectives for this lesson. Use the Learning Objectives to guide your reading, viewing, and thinking.

- Read Chapter 31, "Endocrine Control," pages 506–523, and review page 461 in the Starr textbook.

    Or read, in the Starr/Taggart textbook, Chapter 37, "Endocrine Control," pages 608–625, and review page 547.

- Read the Viewing Notes in this lesson.

View the "Endocrine Control: Systems in Balance" Video Program

After Viewing the Video Program
- Briefly note your answers to the questions at the end of the Viewing Notes.

- Review all reading assignments for this lesson, especially the Chapter 31 Summary on page 522 in the Starr textbook (Chapter 37 Summary on page 624 in the Starr/Taggart textbook) and the Viewing Notes in this lesson.

- Write brief answers to all the Review Questions at the end of Chapter 31 in the Starr textbook (Chapter 37 in the Starr/Taggart textbook) to be certain you understand the text material.

- Complete the Review Activities in this guide to reinforce your understanding of important terms and concepts. Check your answers with the Answer Key and review when necessary.

- Take the Self-Test in this guide to measure your achievement of the Learning Objectives. Check your answers with the Answer Key and review when necessary.

- Complete the Using What You've Learned activities and any other activities and projects assigned by your instructor.

Overview

In the previous lesson, you learned about the system of neural control that enable humans and other organisms to detect and respond to changes in their environment. While this system is both powerful and efficient, all of the activities must occur in close proximity to be effective. In Lesson 22, you will learn about another system which is capable of regulating physiological processes at sites throughout the body: the endocrine system employs unique molecules known as hormones which are transported by the circulatory system to appropriate targets or regions. After much painstaking research, a variety of hormones have been identified from sources such as the pituitary gland, the pancreas, the adrenal glands and others.

As you proceed through the lesson, you will learn about the mechanisms of hormone action on a molecular level. Your study will reveal two major classes of hormones: steroid hormones such as estrogen and cortisol, and peptide hormones such as ADH and insulin. This cast of characters will expand as the true nature of hormone activity unfolds. These characters include not only the hormones, but also secondary messengers, which serve as additional releasing substances and important feedback mechanisms. As you will discover, feedback mechanisms are utilized to fine-tune the hormonal actions and can result in the stimulation or inhibition of further secretions.

In completing this lesson, you will be introduced to the clinical aspects of the endocrine system. Disturbances in hormonal activities have been traced to a variety of disorders such as diabetes, osteoporosis, dwarfism, sterility, and sleeplessness. You'll find that many behavioral problems have also been linked to endocrine disorders and substances that mimic natural hormones. As you become more familiar with the connection between endocrine physiology and behavior, you will gain a greater appreciation of this unique regulatory system.

Learning Objectives

When you have completed all assignments for this lesson, you should be able to:

1. Contrast the structure and general role of an *exocrine* gland with that of an *endocrine* gland.

2. Define a *hormone*, and describe in general how hormones and other *signaling* molecules integrate and control metabolic activities in animals.

3. Compare how the endocrine system and the nervous system control body functions, and describe how these systems interact.

4. Describe the relationship of the vertebrate hypothalamus to the *pituitary* gland, and discuss how this neuroendocrine center controls secretions from the *anterior* and *posterior* pituitary.

5. Describe how homeostatic feedback loops can help maintain normal hormone levels using examples from both pituitary-controlled and pituitary-independent systems.

6. Explain how blood sugar levels are regulated by hormones.

7. Using examples, describe some of the consequences of abnormal endocrine function in the human body.

8. Describe why hormones have an effect on some cells and not others, and compare how *steroid* and *nonsteroid* hormones act on target cells.

9. Describe some methods scientists use to study the endocrine system and the effects of hormones.

Viewing Notes

The video program for Lesson 22 ties together many key concepts of endocrine control. Not only is the endocrine system a primary regulator of homeostasis in the body, it also acts as an indicator of behavioral change. Dr. George Hunt introduces these ideas through his research on the reproductive effects of DDT in California gulls. His findings on the number of eggs found in the nests and the lack of male gulls in the colony has led Dr. Michael Fry to look for a disruption in their delicate hormonal balance. As you will see, a surprising explanation emerges: even at low levels of DDT, its actions mimic the hormone estrogen, causing both feminization of the male gulls and increased egg production by the female gulls. We are left with the idea that humans are also exposed to many substances that may mimic or inhibit hormone action.

The next segment explores the role of endocrine and neural control on the regulation of stress. Some responses to stress today have their origin in the ancient struggle for survival, and are often inappropriate for dealing with immediate or perceived threats in today's world. It is at this point that you'll meet patient Diane Pestolesi, who is seen learning biofeedback techniques in an attempt to reduce the physiological reactions to her environmental stresses and reach a healthier physical and mental state. You'll hear from endocrine researcher Dr. Catherine Rivier, who expounds on this physiological basis of behavior. She will present the basics of pituitary function, effects on the adrenal gland and the feedback loop regulating the reactions to stress. Knowledge of how this system works allows us to devise strategies and treatments to maintain our internal environments even under stress.

In the final segment of the video program, you'll meet radio commentator Pat Gallagher. Gallagher comments on the challenge for him and his daughter Courtney as they try to cope with diabetes. With the help of Professor Gregory Erickson, you'll learn about the production and regulation of insulin and its role in diabetes. You'll also learn about other disease states that can follow the disturbance of glucose metabolism. Returning to Pat Gallagher and his daughter, you'll see how the invention of a pump using human insulin can help to regulate their energy needs and allow them to live the highest quality of life.

When you watch the video program, consider the following questions:

1. Why might seagulls be more susceptible to the effects of DDT than other animals?

2. Why are steroid hormones able to effect changes within the cell more easily than peptide hormones?

3. Why do you think reproductive systems are targeted by disruptions in endocrine activity?

4. Could there be natural substances in the food we eat that also mimic hormone action or inhibition?

5. Do you believe the endocrine system developed as a need to control the internal or external environment? How would you justify your answer as it relates to survival in the wild?

6. What factors, other than stress, can disturb hormone balance?

7. Are the effects of prolonged stress reversible or permanent?
8. How would you measure the success of biofeedback learning?
9. What factors determine the amount of insulin needed at any given time?
10. Can you think of any other treatments or strategies which might alleviate the need for an insulin pump?

Review Activities

Matching

Match the terms listed below with the definitions that follow. Check your answers with the Answer Key and review any terms you missed.

I.

___ 1. target cells
___ 2. hormone
___ 3. pituitary gland
___ 4. endocrine gland
___ 5. pancreatic islets
___ 6. second messengers
___ 7. steroid hormones
___ 8. adrenal medulla
___ 9. gonads
___ 10. peptide hormones

a. endocrine gland associated with the brain and having posterior and anterior lobes for hormone secretion

b. hormones derived from cholesterol which are lipid soluble, binding to receptors inside the cell

c. contains receptors to which hormones bind

d. molecules that complete hormonal activities and may enhance their effects

e. clusters of endocrine cells in the pancreas having alpha, beta, and delta types

f. a signaling molecule secreted into the bloodstream and delivered to target cells and receptors where activity occurs

g. hormones derived from amino acids which are water soluble, binding to receptors on the target cell membrane

h. the inner region of adrenal glands

i. reproductive organs: ovaries and testes; secrete hormones affecting egg and sperm maturation

j. a storage or production gland that secretes hormones

II.

___ 1. ADH   ___ 6. FSH
___ 2. ACTH   ___ 7. TSH
___ 3. PTH   ___ 8. LH
___ 4. cAMP   ___ 9. insulin
___ 5. STH (GH)   ___ 10. melatonin

a. promotes growth, cell division, metabolism, targets most cells

b. targets adrenal cortex to release steroid hormones there

c. increases calcium level in blood while targeting the kidneys and bones

d. stimulates males testosterone release and sperm maturation

e. decreases the glucose level in blood, targets liver, fat, and muscle tissue

f. second messenger capable of enhancing hormone action

g. regulates water retention by targeting the kidneys

h. stimulates female estrogen secretion and egg maturation

i. affects biorhythms and sexual activity, targets gonads and the brain

j. stimulates release of thyroid hormones

k. in females, stimulates progesterone secretion and ovulation

III.

____ 1. hypothalamus
____ 2. feedback loops
____ 3. inhibitors
____ 4. releasers
____ 5. goiter
____ 6. stress reaction
____ 7. rickets
____ 8. diabetes type one
____ 9. hypoglycemia
____ 10. diabetes type two

a. hypothalamic molecules that decrease secretions from the anterior lobe of the pituitary gland

b. beta cells are attacked resulting in a lack of insulin

c. condition of low blood glucose causing alpha cells to release glucagon

d. classic fight-or-flight response or the ability to override the usual regulatory controls for immediate action to occur

e. hypothalamic molecules that stimulate secretions of hormones from the anterior lobe of the pituitary gland

f. beta cells produce less insulin due to lack of response by the target cells

g. primarily a disease of bones due to a vitamin D deficiency and lack of calcium absorption; may involve parathyroid hormone

h. a structure of the brain that produces hormones, some of which are stored in the posterior pituitary gland prior to secretion

i. an enlargement of the thyroid gland due to excess TSH production and insufficient thyroid hormone production

j. increase or decrease of hormonal response due to concentration gradients from the source or the target area affecting further activity

Completion

Fill in each blank with the most appropriate term from the list for that paragraph. A term may be used once, more than once, or not at all. If a question requires two or more answers in succession, they may be in any order (unless the question indicates otherwise). Check your answers with the Answer Key and review when necessary.

1. The pancreas is a _____ that performs many critical functions in the body. Its _____ cells secrete digestive enzymes used daily. The _____ portion contain clusters of cells known as _____. These are composed of three types of _____ secreting cells. The _____ cells secrete somatostatin which helps regulate insulin and glucagon and some digestive functions. The _____ cells secrete glucagon to help raise the blood _____ level. Finally, the _____ cells secrete a hormone which lowers the blood glucose level. This critical hormone is _____. Following a meal, insulin is released targeting liver, fat, and _____ cells. Glucose is taken out of the bloodstream and can now enter these cells for immediate use or be stored as _____ molecules. As the blood glucose _____, insulin secretion slows. Below a certain point, lower blood glucose levels stimulate the alpha cells to secrete _____ which helps to convert glycogen to glucose in the liver. This in turn tends to _____ the blood glucose level back to the set point. This basic regulation is complicated by several other factors including the action of _____.

alpha	glycogen
beta	hormone
decreases	increase
delta	insulin
endocrine	islets
enzyme	kidney
exocrine	lymphatics
gamma	melatonin
gland	muscle
glucagon	somatostatin
glucose	

284   Lesson 22 / Animals: Endocrine Control

2. The connection between the nervous system and the endocrine system is most obvious when looking at the hypothalamus and the _____ gland. The _____ produces ADH and _____, which are stored in the _____ lobe of the pituitary. Later, these hormones are secreted into the _____ fluid and then to the bloodstream. A less direct method of interaction is seen in the _____ lobe of the pituitary. In this case, the lobe _____ the hormones while the hypothalamus directs or triggers the secretion process. The hypothalamus usually sends _____ to stimulate secretion but can also use _____ to decrease or stop hormone secretion. Altogether, _____ hormones are affected by the _____ between the hypothalamus and pituitary.

anterior	pituitary
eight	posterior
hypothalamus	produces
inhibitors	prolactin
interaction	recycles
interstitial	releasers
nine	seven
oxytocin	spinal

3. Hormonal regulation can be disturbed in a variety of ways. _____ mechanisms can fail, causing conditions such as goiter. Goiter results from an excess secretion of _____, leading to an overstimulated _____ gland. In this case, it was determined that a lack of dietary _____ could be causing the problem. Once iodine was available to the thyroid, _____ was reestablished. Classic reactions occur when growth hormone is decreased in childhood which may lead to _____. On the other hand, too much _____ in childhood may lead to _____ or _____ in adulthood. A final example of the sensitivity of endocrine regulation is the action of melatonin. If the production of this hormone in the _____ gland _____, one may become drowsy or _____. As the hormone level _____, more activity and wakefulness results. The interesting aspect of these effects is that melatonin secretion increases with darkness and decreases with _____ stimulation. Thus, environmental regulation of hormone secretion matches with the _____ seen in the natural world.

acromegaly	increases
biorhythms	iodine
calcium	light
decreases	pineal
dwarfism	pituitary
feedback	sleepy
FSH	somatotrophin
gigantism	temperature
goiter	thyroid
homeostasis	TSH

Self-Test

Select the one best answer for each question. Check your answers with the Answer Key and review when necessary.

1. One way in which exocrine glands differ from endocrine glands is that exocrine glands
   a. store many kinds of molecules but do not produce any of them.
   b. secrete their products through ducts.
   c. are only located in the brain.
   d. do not develop until puberty.

2. Hormones can be produced by all of the following EXCEPT
   a. endocrine cells.
   b. neurons.
   c. cartilage.
   d. endocrine glands.

3. The endocrine system differs from the nervous system in its ability to
   a. deliver hormones through the bloodstream to distant targets.
   b. regulate temperature variations in the body.
   c. stimulate metabolic reactions to oxygen consumption.
   d. do all of the above.

4. The hypothalamus produces and controls the secretion of which two hormones associated with the posterior pituitary gland?
   a. ADH and PRL
   b. oxytocin and ADH
   c. STH and oxytocin
   d. PRL and STH

5. An example of a negative feedback loop is
   a. cortisol secretion from the adrenal cortex.
   b. LH surge due to an increase in estrogen.
   c. insulin secretion due to low glucose levels.
   d. none of the above.

6. Blood glucose levels are regulated by
    a. glucagon.
    b. eating habits.
    c. insulin.
    d. all of the above.

7. All of the following are examples of abnormal endocrine system function **EXCEPT**
    a. pituitary dwarfism.
    b. premature onset of puberty in females.
    c. sickle-cell anemia.
    d. acromegaly.

8. Steroid-type hormones act on target cells by
    a. attaching to cilia on the cell surface.
    b. stimulating secondary messengers to action.
    c. converting cholesterol to an enzyme.
    d. interacting with DNA inside the cell's nucleus.

9. Peptide-type hormones act on target cells by
    a. attaching to membrane receptors and initiating action by a secondary messenger.
    b. converting glycopeptides to cAMP.
    c. entering the nucleus as a complex receptor.
    d. responding to hypothalamic signals using peptides as a trigger mechanism.

10. Scientists can study the effects of testicular feminization by
    a. monitoring the development of secondary sexual characteristics.
    b. injecting molecules that mimic estrogen and monitoring the effects.
    c. injecting an excess of testosterone and monitoring the results.
    d. all of the above.

Using What You've Learned

Based on your own interests or your instructor's requirements, complete one or more of the following activities.

1. Arrange to use a portable glucose monitoring device. Test your glucose level before breakfast and compare with the expected range for fasting glucose. Are you above or below this range?

2. A classic diagnostic test for diabetes or hypoglycemia is the glucose tolerance test. Ranging from two to six hours, the patient's ability to assimilate a given amount of glucose is determined. Using library references, compare the resulting "curves" for diabetics, hypoglycemics and normal patients. What conclusions can you make?

3. Psychobiologists attempt to discover the physiological basis of behaviors. Working backwards instead, make a list of possible behaviors you might observe due to the actions of various hormones.

4. Obtain the insert from an over-the-counter melatonin package. Is this a better alternative than using traditional "sleeping pills"? What are the side effects?

5. Interview a female subject willing to talk about her experiences during menopause. Attempt to match these symptoms with changes in the endocrine system during this time frame.

6. Investigate biofeedback techniques and how they may be used to regulate the endocrine system. How do biofeedback devices work? What are the medical or psychiatric uses of biofeedback?

Challenge Questions

1. Explain the role of oxytocin in the birth process. What triggers the hormone release at that time?

2. Was the development of the endocrine system in response to a lack of hormone-like substances in the diet or environment?

3. Since a very small amount of hormone or hormone-like substance is necessary to cause a physiological response, describe a possible "hormone warfare" plan to regulate or control the population.

4. As we begin to see females giving birth after age 50, what implications does this have for the family members and society as a whole?

5. Would hormones be more efficient if they could all be converted to a lipid soluble state?

6. Why is the hormone ecdysone important for insects?

7. What effects do plant hormones have on the human body?

LESSON 23

Animals: Reproduction and Development

Assignments

For the most effective study of this lesson, we suggest that you complete the assignments in the sequence listed below:

Before Viewing the Video Program
- Read the Overview and Learning Objectives for this lesson. Use the Learning Objectives to guide your reading, viewing, and thinking.

- Read Chapter 38, "Reproduction and Development," pages 630–668, in the Starr textbook.

   Or read, in the Starr/Taggart textbook, Chapter 44, "Principles of Reproduction and Development," pages 744–761, and Chapter 45, "Human Reproduction and Development," pages 762–790.

- Read the Viewing Notes in this lesson.

View the "Animal Reproduction and Development" Video Program

After Viewing the Video Program
- Briefly note your answers to the questions at the end of the Viewing Notes.

- Review all reading assignments for this lesson, especially the Chapter 38 Summary on pages 666–667 in the Starr textbook (Chapter 44 and Chapter 45 Summaries on pages 770 and 798 respectively in the Starr/Taggart textbook) and the Viewing Notes in this lesson.

- Write brief answers to all the Review Questions at the end of Chapter 38 in the Starr textbook (Chapters 44 and 45 in the Starr/Taggart textbook) to be certain you understand the text material.

- Complete the Review Activities in this guide to reinforce your understanding of important terms and concepts. Check your answers with the Answer Key and review when necessary.

- Take the Self-Test in this guide to measure your achievement of the Learning Objectives. Check your answers with the Answer Key and review when necessary.

- Complete the Using What You've Learned activities and any other activities and projects assigned by your instructor.

Overview

In previous lessons, you've seen how individual cells and primitive multicellular organisms perpetuate themselves through the reproductive process. In this lesson, you'll explore the reproductive systems and processes used by animals higher up on the evolutionary ladder, including *Homo sapiens*.

You'll begin by learning about the adaptive advantages of sexual reproduction and how it compares to asexual reproduction in other organisms. Next, you'll be given a visual and narrative tour of embryonic development, beginning with the fertilization of an egg by a sperm cell and continuing through the processes of cellular division, differentiation, and migration that characterize the developing embryo. By observing similarities in the embryos of different species, you'll gain a greater appreciation for the common lineage we share with other members of the animal kingdom. You'll also learn at what point the human embryo begins to exhibit the morphological characteristics that distinguish it as a species and as a unique life form.

From this general overview of embryonic development, the lesson will orient you to the unique structures and hormonal processes of the male and female reproductive systems. During this orientation, you'll learn how sperm cells are produced in males and how egg cells mature during the female menstrual cycle. You'll review in some detail the events that occur when a mature egg fails to become fertilized (menstruation) or becomes fertilized by a sperm cell as a result of sexual intercourse (pregnancy).

From here, Lesson 23 follows the development of the human embryo and fetus through the three trimesters of pregnancy. You'll learn about the function of the placenta and how the fetus is supported during its growth. You'll also learn at which stages of pregnancy characteristic traits and behaviors arise.

Lesson 23 will also survey some of the more common disorders of the reproductive system, including sexually transmitted diseases, infertility, and the effects of infection and drug use on pregnancy. You'll benefit from current research on the nutritional requirements for pregnancy and ways to reduce the common effects of aging.

Learning Objectives

When you have completed all assignments for this lesson, you should be able to:

1. Compare *asexual* and *sexual* reproduction, and discuss the adaptive advantages and problems associated with having separate sexes.

2. Describe the key mechanisms of embryonic animal development, including *determination*, *differentiation*, and *morphogenesis*.

3. Describe the structure and function of the human male and female *reproductive systems*.

4. Describe the events of a typical human *menstrual cycle*, including hormones secreted from the pituitary and ovary, and the resulting physical changes at the ovary, uterus, and mammary glands.

5. Outline the principal events of prenatal human development, noting the major events of each *trimester*.

6. Identify the changes that occur in an individual from birth to death as a continuation of the development and aging processes, and note the possible causes of aging.

7. Discuss methods that can be used to control human fertility, and describe some modern techniques used to help couples who cannot have children through normal means.

8. Discuss the importance of the mother's nutrition on fetal development, and identify the risks of taking drugs during pregnancy.

9. Describe some methods that scientists use to study animal reproduction and development.

Viewing Notes

Using a combination of case studies, interviews, microscopic images and computer simulation, the program for Lesson 23 will reveal secrets of the human reproductive process.

In the first part of the program, the husband-and-wife team of Dr. Marianne Bronner-Fraser and Dr. Scott Fraser share their research in embryonic development. They have been studying developing embryos of different species to answer a question that has perplexed scientists for some time: How does a simple fertilized egg differentiate and mature into a complex multicellular organism?

Through videomicrography, you'll witness the intricate processes that transform one cell into two cells, then four cells, and eventually into a fluid-filled ball of cells called a blastula. It is at this point that cell differentiation takes the developing embryo down a path leading to a unique expression of a particular species.

In Part Two, the theme of "Fertility" provides a springboard for a discussion of the human reproductive process. You'll follow the case of a young married couple who encountered problems with natural pregnancy and the fertility specialist who was able to help them through in vitro fertilization. As the story unfolds, you'll learn about some of the factors that promote and interfere with fertility in both males and females.

Through computer animation, you'll follow the events that characterize the menstrual cycle, and learn what happens at the microscopic level when the female egg is left unfertilized or is fertilized prior to pregnancy. Using visuals rarely seen by the naked eye, Part Three will show you the miracle of life as it proceeds from the growth of a human embryo to the birth of a fully developed fetus.

Part Three will also present some of the more important lifestyle considerations for women considering pregnancy, including nutrition, exercise and exposure to substances such as alcohol and nicotine. You'll meet Dr. Peggy Grau, an obstetrician with many years of experience helping new mothers prepare themselves for giving birth. By explaining the course of events in the developing fetus, Dr. Grau will show you how harmful substances consumed by the mother can affect the growth and health of the fetus prior to birth.

As you watch the video program, consider the following questions:

1. How does asexual reproduction differ from sexual reproduction?
2. What are the advantages of sexual reproduction in terms of the survival of a species?
3. What happens to a fertilized egg cell during the early stages of embryonic development?
4. Why is it important for cells to differentiate and interact during embryonic development?
5. What kinds of male and female factors influence human fertility?

6. Where and how are sperm cells produced?
7. What are the sequence of events of the female menstrual cycle?
8. What is *in vitro fertilization* and how does it work?
9. What is the sequence of events in the development of a human fetus?
10. What kinds of lifestyle factors affect pregnancy and the developing fetus?

Review Activities

Matching

Match the terms listed below with the definitions that follow. Check your answers with the Answer Key and review any terms you missed.

I.

___ 1. corpus luteum
___ 2. differentiation
___ 3. endometrium
___ 4. estrogen
___ 5. fertilization
___ 6. implantation
___ 7. menstrual cycle
___ 8. morphogenesis
___ 9. ovary
___ 10. placenta
___ 11. progesterone
___ 12. testes
___ 13. testosterone
___ 14. trimester

a. the primary female reproductive organ in which eggs develop

b. the fusion of the nuclei of a sperm and an egg

c. the primary male reproductive organ where sperm are formed

d. the periodic release of oocytes and priming of the uterine lining to receive a fertilized egg

e. a glandular structure that secretes hormones that maintain the lining of the uterus

f. one of the three periods that comprise pregnancy

g. processes by which differentiated cells organize into tissues and organs

h. the inner lining of the uterus

i. the tissues and extraembryonic membranes that exchanges nutrients and waste materials between the mother and fetus

j. a female sex hormone that helps oocytes mature and influences changes in the uterine lining

k. male hormone that helps control male reproductive functions

l. selective activation or suppression of some of a cell's genes (or DNA) to allow only certain proteins to be synthesized for specialized functions

m. female sex hormone secreted by the ovaries

n. the attachment of a blastocyst to the uterine lining

Lesson 23 / Animals: Reproduction and Development   295

II.

___ 1. allantois ___ 7. embryo
___ 2. amnion ___ 8. endoderm
___ 3. blastula ___ 9. gastrulation
___ 4. chorion ___ 10. mesoderm
___ 5. cleavage ___ 11. oocyte
___ 6. ectoderm ___ 12. yolk sac

a. the stage of embryonic development in which cells form two or three tissue layers

b. the outer primary tissue layer that develops into the outer layer of the integument and tissues of the nervous system

c. an extraembryonic membrane that becomes the fluid-filled sac in which the embryo grows

d. an extraembryonic membrane that becomes a major part of the placenta

e. the intermediate primary tissue layer that develops into muscles, internal skeleton, connective tissue, and many internal organs

f. an extraembryonic membrane that forms during embryonic development and assists with respiration and storage of metabolic wastes

g. an extraembryonic membrane that becomes the site of blood cell formation in humans

h. embryonic stage characterized by the mitotic division of a zygote to form a ball of cells

i. embryonic stage characterized by a hollow ball of cells created by cleavage

j. an immature egg

k. an animal during its early stage of development; characterized by cleavage and tissue differentiation

l. the inner primary tissue layer that develops into the inner lining of the gut and associated organs

Completion

Fill in each blank with the most appropriate term from the list for that paragraph. A term may be used once, more than once, or not at all. If a question requires two or more answers in succession, they may be in any order (unless the question indicates otherwise). Check your answers with the Answer Key and review when necessary.

1. In humans, sexual reproduction begins with the formation of _____ in the male _____ and _____ in the female _____. In the male, the production of _____ is influenced by hormones such as _____ and _____ secreted by the anterior _____ gland. In the female, secretions of _____, _____, _____, and _____ stimulate changes in _____ and the _____. During the female menstrual cycle, a follicle matures in the _____ as the _____ thickens. A surge of _____ triggers _____, and a _____ forms from the remnants of the follicle. The corpus luteum secretes _____ and _____, priming the endometrium for _____. If fertilization occurs, the corpus luteum is maintained.

LH	progesterone
FSH	embryo
ovary	testes
eggs	sperm
endometrium	testosterone
ovulation	estrogen
placenta	ovaries
pituitary	thymus
corpus luteum	implantation
fertilization	

2. Embryonic development in animals begins when _____ and _____ develop in the female and male reproductive organs respectively. _____ brings the sperm and egg together to form a _____. _____ enables the zygote to become a multicellular _____, which becomes surrounded by four extraembryonic membranes (_____, _____, _____, and _____). The cells of the embryo further differentiate into primary tissue layers (_____, _____, and _____). Through the process of _____, specialized tissues and organs begin to form and mature.

yolk sac	zygote
endoderm	differentiation
mesoderm	ectoderm
gastrulation	chorion
amnion	cleavage
fertilization	gamete
sperm	allantois
fetus	ovulation
oocytes	blastocyte
morphogenesis	organs
embryo	

3. Through deliberate or accidental intervention, humans can either promote or inhibit the reproductive process. _____ is a method frequently used to promote conception when parents are unable to conceive through natural methods. Birth control methods range from the most effective (_____, _____, _____) to moderately effective (_____, _____, _____). Embryonic development can be compromised by _____ crossing the _____. Substances such as _____, _____, and _____ may stunt growth, impair mental development, or cause birth defects.

placenta	abstinence
in vitro fertilization	douching
abstinence	rhythm method
teratogens	artificial conception
pathogens	diaphragms
tobacco	alcohol
condoms	endometrium
vasectomy	tubal ligation
induced insemination	pills
withdrawal	drugs

Self-Test

Select the one best answer for each question. Check your answers with the Answer Key and review when necessary.

1. The principal advantage of sexual reproduction over asexual reproduction is that sexual reproduction
    a. ensures that all offspring will survive under current environmental conditions.
    b. provides both an emotional and a physical release from tension.
    c. requires sperm and egg production to be synchronized.
    d. produces new allele combinations that may confer a selective advantage.

2. Morphogenesis describes the process whereby
    a. cells rearrange to form the linings of the body cavity.
    b. dividing cells form into a ball surrounding a fluid-filled space.
    c. cells begin to specialize in the production of certain kinds of proteins.
    d. cell growth and migration lead to organ and tissue organization.

3. Which of the following is the site of sperm cell production in the male reproductive system?
    a. testes
    b. seminal vesicle
    c. prostate gland
    d. corpus callosum

4. In the female reproduction system, the chamber in which the embryo develops is the
    a. fallopian tube.
    b. ovary.
    c. uterus.
    d. vagina.

5. During the early phase of the menstrual cycle, secretions of FSH and LH from the pituitary gland
    a. stimulate the growth of a follicle in the ovary.
    b. cause an oocyte to be released into the fallopian tube.
    c. trigger the secretion of estrogen from the thalamus.
    d. prepare the endometrium for pregnancy.

6. Near the end of the second trimester of human development,
   a. limbs and digits are beginning to form.
   b. the eyelids are open.
   c. head growth surpasses that of all other body regions.
   d. the fetus is ready to be born.

7. Which of the following explains the aging process?
   a. Increased susceptibility to autoimmune disease
   b. Limitation in the number of times a cell can divide
   c. Loss of cell's ability to repair itself
   d. No one knows what causes aging.

8. Which of the following is the most effective form of birth control?
   a. birth control pills
   b. douching
   c. rhythm method
   d. tubal ligation

9. Alcohol consumed by the mother during pregnancy
   a. is normally filtered out by the placenta.
   b. can be harmful to the fetus when consumed in excess.
   c. will result in a baby born with fetal alcohol syndrome.
   d. actually enhances fetal digestion of some nutrients.

10. By studying embryonic development in different species, scientists hope to determine
    a. the mechanisms of the aging process.
    b. how the physical characteristics of the fetus can be manipulated.
    c. how the growth process can be interrupted at different stages.
    d. how a single cell can develop into a complex, multicelled organism.

Using What You've Learned

Based on your own interests or your instructor's requirements, complete one or more of the following activities.

1. Identify a family member or friend who has recently had a baby. Ask to see any sonograms taken during the pregnancy, and identify the stage of the pregnancy during which each sonogram was taken. What fetal structures can you identify?

2. If possible, visit a sperm bank and interview the staff to find out how such a facility operates. Or visit a fertility clinic and investigate the various methods available to promote pregnancy among infertile couples.

3. Visit the library and scan recent science publications for current research on aging. In addition to those described in your text, what other theories have been postulated to explain the aging process? How might this information be useful in terms of slowing or stopping the aging process in the future?

4. Use the library to investigate the underlying causes of sex crimes such as child molestation, statutory rape, and the like. What kinds of treatments or therapies are available to help people with these inclinations? What is the success of these treatments or therapies?

5. Research a common male or female sexual dysfunction and report on its cause(s), clinical signs and symptoms, and recommended course of treatment. What risk factors, if any, make individuals susceptible to this dysfunction? How successful are current modes of treatment?

Challenge Questions

1. If sexual reproduction is advantageous to most plants and animals, why does asexual reproduction still persist? What possible advantages would asexual reproduction have over sexual reproduction?

2. How do fraternal twins develop? What produces identical twins? Why do nonidentical twins show more genetic variability than identical twins?

3. Why is it that only one sperm can fertilize an egg? What stops other sperm from fertilizing the same egg?

4. Why is it that the risk of birth defects increases with mothers over 40? Does the age of the male make any difference in terms of successful fertilization?

5. What organs does the placenta supplement or replace?

6. At what point in embryonic development do cells begin to differentiate? What are the first kinds of differentiated cells formed?

7. What causes cells to migrate during embryonic development to form the blastula, various axes and linings, and eventually the vital organs and tissues of the fetus?

8. What kinds of factors affect fertility? What measures are available to improve fertility?

9. In what ways does the female menstrual cycle provide a good example of positive or negative feedback inhibition?

10. Do men experience a type of "menopause" as they get older? If so, what characterizes this transition?

LESSON 24

Populations and Communities

Assignments

For the most effective study of this lesson, we suggest that you complete the assignments in the sequence listed below:

Before Viewing the Video Program

- Read the Overview and Learning Objectives for this lesson. Use the Learning Objectives to guide your reading, viewing, and thinking.

- Read Chapter 39, "Population Ecology," pages 670–687, and Chapter 40, "Community Interactions," pages 688–705, in the Starr textbook.

  Or read, in the Starr/Taggart textbook, Chapter 46, "Population Ecology," pages 792–811, and Chapter 47, "Community Interactions," pages 812–833.

- Read the Viewing Notes in this lesson.

View the "Populations and Communities" Video Program
After Viewing the Video Program

- Briefly note your answers to the questions at the end of the Viewing Notes.

- Review all reading assignments for this lesson, especially the Chapter 39 and Chapter 40 Summaries on page 686 and 704 in the Starr textbook (Chapter 46 and Chapter 47 Summaries on page 820 and 842 respectively in the Starr/Taggart textbook) and the Viewing Notes in this lesson.

- Write brief answers to all the Review Questions at the end of Chapters 39 and 40 in the Starr textbook (Chapters 46 and 47 in the Starr/Taggart textbook) to be certain you understand the text material.

- Complete the Review Activities in this guide to reinforce your understanding of important terms and concepts. Check your answers with the Answer Key and review when necessary.

- Take the Self-Test in this guide to measure your achievement of the Learning Objectives. Check your answers with the Answer Key and review when necessary.

- Complete the Using What You've Learned activities and any other activities and projects assigned by your instructor.

Overview

An introductory course in biology would not be complete without presenting a broader picture of our living world. Throughout the previous sections, lessons have explained concepts ranging from cellular function and inheritance to the principles of evolution and diversity. When these ideas were applied to the plants and animals, complexities of their structure and function were dramatically demonstrated. In Lesson 24, we will begin to explore how the organisms of our biological world interact with each other and their unique environments.

The lesson begins by defining the principles regulating populations of organisms. Patterns of growth and their mathematical relationships are introduced to help explain the prevalence or scarcity of the various species we see in nature. Exponential and logistic growth models are defined as well as modifying factors such as limited resources and competition. Life tables and survivorship curves are also shown to demonstrate patterns typical of most populations.

Those principles and tools will provide a foundation for understanding the dilemma of a rapidly expanding human race. The lesson will challenge you to evaluate various projections of human population growth in the future. After presenting the environmental, economic, and social factors affecting its growth, a warning is issued to address the problem of overpopulation now or face the limits nature will force upon a population above its carrying capacity.

As the lesson continues, you will learn about communities of populations and the habitats that define them. Interactions between various species can have a full range of effects or none at all. Examples of mutualism, competition, predation, and parasitism demonstrate the forces at work within the community. Ultimately, the ability to survive, whether by camouflage, mimicry or other behaviors, is a continuous struggle as habitats and niches within them constantly change.

The lesson concludes with a global look at the biodiversity of communities. This discussion leads naturally into the next lesson and to new levels of complexity—ecosystems and the biosphere.

Learning Objectives

When you have completed all assignments for this lesson, you should be able to:

1. Define the terms *population*, *community*, *ecosystem*, and *biosphere*, and note how they relate to each other.

2. Use a *growth curve* to describe the changes in an isolated bacteria population, and explain the cause for each stage of growth: lag, exponential growth, maximum stationary, and death.

3. Compare the isolated bacteria growth curve with the idealized *S-shaped* curve that many populations follow in nature, and discuss reasons for their similarities and differences.

4. Describe how *density-dependent* and *density-independent* controls might modify an S-shaped growth curve.

5. Describe the human growth curve, noting possible reasons for periods of growth and stability, and discuss where this curve may be headed in the future.

6. Note the relationship of birth rate and death rate to changes in total population size, and describe the basic principles of the *demographic transition model*.

7. Use an *age structure diagram* to determine how a population is changing in size, and discuss the reasons that a population will continue to grow for many decades even beyond the attainment of its *replacement rate*.

8. Identify and distinguish between different types of species interaction, noting the direct effect of each interaction on the species involved.

9. Define and give examples of *competitive exclusion*, and relate this principle to the diversification of species into different *niches*.

10. Describe how communities develop and diversify, noting the roles of *succession* and *disturbances*.

Viewing Notes

Until now, the video programs you have watched have focused upon individual organisms and their unique characteristics. We now attach a wide angle lens to our camera in order to gain a larger view of how these organisms interact within their own group setting and with other species in the same environment. As you will see, the basic concepts of ecology are introduced through the study of various populations and communities.

The adventure begins with Dr. Jack Burk explaining the fragile existence of an endangered plant species, the Santa Ana River Woolly Star. You will discover how the survival of such sensitive species can indicate the general stability of any given habitat. Through this example, the importance of population growth curves and their relationship to the environment is discussed.

The next segment takes you on an owl hunt with biologist Pete Bloom and Dr. Allan Schoenherr at the Starr Ranch Audubon Sanctuary. The interactions between species are introduced through the predator-prey relationship. Examples of symbiotic relationships are presented and documented in dramatic fashion. By studying the habits of various species, more insight can be gained about the effects of various changes on the community as a whole.

The final part of this video is concerned with the controversial issues of our own human population. Ranging from our social history to advanced technology, reasons are presented for the growth and population trends we see today. You will have the opportunity to conceptualize future population growth by studying an age structure diagram and its implications. The video closes with some challenging thoughts relating the effects of human population growth to the health of the entire planet.

As you watch the program, consider the following questions:

1. If an endangered species such as the Santa Ana River Woolly Star were to become extinct, what effects might this have on other species?
2. What do the J- and S-shaped population curves indicate about the life span of an organism?
3. Does competitive exclusion limit the diversity within a given community?
4. Why are field studies, such as the monitoring of owls, important in the study of ecology?
5. How does mutualism differ from commensalism?
6. What effect did the agricultural revolution have on the growth of the human population?
7. What does the demographic transition model say about birth and death rates?
8. In your opinion, is a zero population growth rate possible?
9. What critical factors could affect the age structure diagram?
10. Do you believe that the human population can stabilize without widespread starvation, war, or epidemics?

Review Activities

Matching

Match the terms listed below with the definitions that follow. Check your answers with the Answer Key and review any terms you missed.

I.

___ 1. ecology
___ 2. population
___ 3. ecosystem
___ 4. community
___ 5. biosphere
___ 6. carrying capacity
___ 7. exponential growth
___ 8. logistic growth
___ 9. S-shaped curve
___ 10. J-shaped curve

a. the maximum number of organisms that can be maintained indefinitely within a specific environment

b. a group of organisms of the same species living within a given area or habitat

c. the entire area of earth, water, and air which sustains life on the planet

d. displays a pattern of logistic population growth

e. consists of communities of organisms and their abiotic environment interacting by exchanging and recycling energy

f. the size of a population expands by ever-increasing increments during successive (equal) time intervals

g. the branch of science concerned with the interaction of various organisms with each other and their physical and chemical environment

h. growth pattern seen from low-density slow growth to rapid increase to a leveling rate when the carrying capacity is achieved

i. displays a pattern of exponential growth

j. a group of populations consisting of all species living within a given area or habitat

II.

____ 1. limiting factor ____ 5. habitat
____ 2. demographic transition model ____ 6. niche
____ 3. survivorship curve ____ 7. climax community
____ 4. zero population growth rate ____ 8. succession community

a. area where a specific organism or population lives, defined by physical, biological, and chemical features

b. orderly changes in a community until a steady state is established

c. human growth rate pattern related to stages of economic development

d. human growth rate pattern when births are equal to deaths over a given time period

e. environmental and biological conditions under which a given organism or species can be sustained

f. necessary resource or essential factor in short supply

g. steady state of species and habitat following a period of change

h. indicates age-specific survival of a population in a given habitat from birth until the last individual dies

III.

____ 1. competitive exclusion ____ 5. interspecific competition
____ 2. parasitism ____ 6. commensalism
____ 3. predation ____ 7. resource partitioning
____ 4. mimicry ____ 8. mutualism

a. a complex evolutionary adaptation allowing for greater survival from predation

b. two different species requiring the same resources cannot coexist in the same habitat forever

c. competing species share resources by utilizing them in different ways, at different times or in specific areas

d. one species harms or destroys the other non-host species

e. competition for resources among different species sharing a habitat

f. one species harms or destroys the other host species

g. both species benefit from their interaction, directly, or indirectly

h. one species benefits while the other is largely unaffected by the interaction

Completion

Fill in each blank with the most appropriate term from the list for that paragraph. A term may be used once, more than once, or not at all. If a question requires two or more answers in succession, they may be in any order (unless the question indicates otherwise). Check your answers with the Answer Key and review when necessary.

1. Most populations have several common characteristics including a _____ distribution pattern and an _____ structure. Growth over time can be _____ when the size is dependent only upon a rate and the current _____ base. This equation, defined by the terms $G =$ _____, can be graphed as population size versus time in a classic _____ curve. In the real world, limiting factors such as _____ and _____ create a maximum _____ for a specific habitat. Growth over time is now denoted as _____ in nature and is expressed as $G =$ _____. When plotted as before, a classic _____ curve is shown. This growth model is _____ although factors other than the _____ are also important. There are still more factors affecting population growth which are _____. Some of these include _____, widespread flooding, and _____.

    reproductive   geometric
    food   exponential
    oil   random
    living space   density dependent
    carrying capacity   density independent
    biological control   birth rates
    S-shaped   logistical
    J-shaped   drought
    clumped   volcanic eruptions
    rN age
    $r_{max} N [(K - N)/K]$

2. The _____ model uses economic development to explain the changes in population growth rates. The stages of economic development begin with the _____ stage when the growth rate is _____. Next, the _____ stage shows a sharp _____ in growth rate. The growth rate slows during the _____ stage and achieves _____ growth in the _____ stage. An example of a country in the transitional stage is _____, while _____ is considered to be in the industrial stage.

   low   preindustrial
   increase   industrial
   decrease   postindustrial
   stability   Brazil
   transition   New Guinea
   population control   France
   zero population   Greenland
   demographic transition

3. Species can interact in many ways causing a positive, negative, or _____ effect upon each other. When both species benefit, this is known as _____, while _____ benefits one species, but harms the host species. _____ occurs when one species is benefited while the other is not affected. The previous three interactions depend on a close relationship or _____ during all or part of their _____ cycle. Other forms of interaction include _____ when only one species benefits and a non-host species is harmed; and _____ when both species are harmed to some extent. Many of these interactions help to _____ populations, define their specific _____, and to provide for stability within the _____.

   species   regulate
   life   community
   balance   biosphere
   neutral   parasitism
   interspecific competition symbiosis
   intraspecific competition predation
   niche   mutualism
   commensalism   mimicry

310   Lesson 24 / Populations and Communities

Self-Test

Select the one best answer for each question. Check your answers with the Answer Key and review when necessary.

1. Which of the following is the proper order of terms describing the most complex to the least complex?
   a. population, ecosystem, community, biosphere
   b. population, community, ecosystem, biosphere
   c. ecosystem, population, community, biosphere
   d. biosphere, ecosystem, community, population

2. An isolated bacteria population experiences stages including the lag, exponential, stationary, and death phases. Which of the following is not a limiting factor in its population growth?
   a. available nutrients
   b. outside humidity
   c. time of incubation
   d. accumulated waste products

3. An S-shaped curve is only an approximation because
   a. carrying capacities are never achieved.
   b. predators do not exist for all species.
   c. conditions can change carrying capacities.
   d. limiting factors do not affect entire populations.

4. Which of the events listed below involve density independent controls capable of modifying an S-shaped curve?
   a. a severe regional hurricane
   b. radioactive fallout from a nearby reactor
   c. fire covering the entire habitat
   d. all of the above

5. The most important factor in slowing human population growth is
   a. emigration rate.
   b. death rate.
   c. birth rate.
   d. immigration rate.

6. Choose the correct statement concerning the demographic transition model:
   a. Population growth declines during the preindustrial stage.
   b. Population size slowly decreases during the postindustrial stage.
   c. Zero population growth is finally achieved during the transitional stage.
   d. The industrial stage is marked by a dramatic increase in population growth.

7. Age structure diagrams may indicate future growth by
   a. a narrow base in the female age groups under 15.
   b. a wide base in the female age groups over 45.
   c. a wide base in the female age groups under 15.
   d. a wide base in the male age groups under 30.

8. A type of species interaction which results in only positive effects is known as
   a. mutualism.
   b. commensalism.
   c. interspecific competition.
   d. neutralism.

9. Competitive exclusion may result in
   a. different species forming unique and overlapping niches in order to survive.
   b. a balanced population, stable into the future.
   c. one species dominating and ultimately eliminating the others from the habitat.
   d. the conditions in both a and c.

10. Succession models indicate all the following EXCEPT
    a. keystone species cannot permanently affect habitats.
    b. the climax community contains species in equilibrium with the environment and each other.
    c. disturbances in the community may be permanently reflected by changes in the succession dynamics.
    d. regional forces of succession may produce similar climax stages.

Using What You've Learned

Based on your own interests or your instructor's requirements, complete one or more of the following activities.

1. Describe your own unique niche in biological and physical terms. Include your interactions with family, friends, work, and school to complete this activity. How does your niche compare with those of animals living in the wild?

2. Building on your text reading, obtain recent census data for California and plot an age structure diagram. With your knowledge and any additional research, comment on possible implications the diagram has for future population growth.

3. Obtain a list of plants native to your area and compare with plants seen in your neighborhood and in local parks. What are some of the effects these new species may have had within the community?

4. Visit a museum of natural history, botanical gardens, or zoo to learn more about the diversity of animals, plants, and their interactions. Use these experiences to become more aware of the natural world and to continue your lifetime study of biology.

5. As a small research project, conduct an informal poll asking the following question: Assuming you could choose the number of children you would have in your family today, how many would there be?

6. Ask at least ten people of both genders and assume these answers stayed the same for the next two generations as well. Calculate the percent increase in population from the second to the third generation and comment on your results.

Challenge Questions

1. Why are many brightly colored organisms likely to be poisonous, bad tasting, or dangerous to other species?

2. Why are parasite-host relationships not considered to be predator-prey relationships?

3. If you are attempting to introduce a new species to your community, what factors would you consider before bringing it there?

4. What is your estimate of the ultimate carrying capacity of the human population? Justify your answer.

5. A new island has been built up above sea level but is 1,000 miles from the nearest land. What sources are available to populate this island with both plant and animal species? What are some of the limiting factors relating to the biodiversity on the island after many years?

LESSON 25

Ecosystems and the Biosphere

Assignments

For the most effective study of this lesson, we suggest that you complete the assignments in the sequence listed below:

Before Viewing the Video Program

- Read the Overview and Learning Objectives for this lesson. Use the Learning Objectives to guide your reading, viewing, and thinking.

- Read Chapter 41, "Ecosystems," pages 706–723, and Chapter 42, "The Biosphere," pages 724–751, in the Starr textbook.

    Or read, in the Starr/Taggart textbook, Chapter 48, "Ecosystems," pages 834–853, and Chapter 49, "The Biosphere," pages 854–881.

- Read the Viewing Notes in this lesson.

View the "Ecosystems and the Biosphere" Video Program

After Viewing the Video Program

- Briefly note your answers to the questions at the end of the Viewing Notes.

- Review all reading assignments for this lesson, especially the Chapter 41 and Chapter 42 Summaries on pages 722 and 750 in the Starr textbook (Chapter 48 and Chapter 49 Summaries on pages 862 and 890 respectively in the Starr/Taggart textbook) and the Viewing Notes in this lesson.

- Write brief answers to all the Review Questions at the end of Chapters 41 and 42 in the Starr textbook (Chapters 48 and 49 in the Starr/Taggart textbook) to be certain you understand the text material.

- Complete the Review Activities in this guide to reinforce your understanding of important terms and concepts. Check your answers with the Answer Key and review when necessary.

- Take the Self-Test in this guide to measure your achievement of the Learning Objectives. Check your answers with the Answer Key and review when necessary.

- Complete the Using What You've Learned activities and any other activities and projects assigned by your instructor.

Overview

In the previous lesson you found out how individual organisms combine to form populations and exhibit unique interactions within their own communities. The lesson also revealed that biodiversity results from a variety of adaptations within specific environments. While this diversity is indeed impressive, the differences seen in each system appear to be balanced by similarities of structure and function. Taking this idea one step further, we can now define the characteristics of ecosystems, not only in terms of the organisms and living conditions but also in terms of how energy is used and recycled. In Lesson 25, you will be introduced to food webs, energy pyramids and biogeochemical cycles and their effects on the ecosystem. Disturbances to the ecosystem, whether by natural means or human interference, can be predicted and monitored using statistical analyses and other quantitative models. As you will see, the long-term effects of such disturbances can be devastating to an ecosystem.

Your exploration of ecosystems continues with a more global picture of the biosphere, which includes the hydrosphere, the lithosphere and the atmosphere. Such forces as the climate, ocean currents and temperature zones help to carve out the distinct biogeographic realms we see on the planet today. These realms are subdivided into large regions of similar vegetation and soil known as biomes. You will explore each of the major biomes, traveling from rain forests to the desert to the Arctic tundra. The focus will then turn to aquatic ecosystems and coastal environments, including estuaries and coral reefs.

As you near the end of Lesson 25, the true character of the biosphere will begin to take shape. An interesting phenomenon called the "El Niño" effect is presented to show how each part of the biosphere is affected by the other. Finally, you will come to the inevitable conclusion that while it may be easier to study the various organisms in isolation, the interrelationships between the living world and the environment as a whole define the Earth as we know it today.

Learning Objectives

When you have completed all assignments for this lesson, you should be able to:

1. Describe the general pattern of energy flow through an ecosystem, and explain how an ecosystem's nutrients are cycled between major groups of organisms.

2. Starting with *producers*, describe the feeding relationships and energy flow observed in ecosystems, using concepts such as *trophic levels*, *food chains* and *webs*, and *ecological pyramids*.

3. Describe the processes at work in *biogeochemical cycles*. Illustrate how organisms interact with chemical reservoirs in the environment, discussing the potential for deficits or excesses in various parts of each cycle.

4. Use an example to illustrate the concept of *biological magnification*.

5. Discuss how the different motions and physical characteristics of the earth can affect climate on a global or regional level.

6. Define a *biome*, and describe how climate, topography, and other physical characteristics cause the unique plant and animal associations found in each region.

7. Describe the physical and chemical factors that influence *zonation* in water-based ecosystems.

8. Describe some of the methods scientists use to study ecosystems and the biosphere.

Viewing Notes

The video program for Lesson 25 vividly illustrates how the survival of living organisms is dependent upon interactions with the non-living world. Part One discusses the concept of trophic levels and food webs as a way to study the effects of changes in various habitats. U.S. Fish and Wildlife biologist Dick Zembal explains the need for preserving ecosystems in order to prevent ripple effects throughout the food web. Using the Light-Footed Clapper Rail, an endangered bird species, as an example, Zembal describes how changes at lower levels of the ecosystem can be magnified at a higher level, threatening the very existence of these specialized birds.

While the complexity of these interactions may seem overwhelming at first, you will be able to observe how scientists break these interactions down into a series of more easily understood events. An example of this is shown in the second segment, "Fire in the Forest," in which wildlife biologist Diane Freeman describes the different trophic levels found in ecosystems. You'll discover how forest fires, giving the appearance of complete destruction, actually help to recycle the nutrients essential for the survival of the overall ecosystem. Because the need for periodic renewal in the forest is now understood, fires are deliberately set to enable the process of biogeochemical recycling to occur. Many other examples of energy flow and cycling present in nature are discussed.

A more global view awaits you in the final segment of the program. You'll begin with Dr. Warren Blier of UCLA's School of Atmospheric Science, introducing the ideas of regional weather patterns and the resulting forecasts. The interaction of the living world with long-term climate patterns has helped to define the biomes of the Earth. Dr. Blier continues to illustrate our interactions with the physical world, explaining the variations in weather patterns in different regions. Next, biologist Mark Poth of the U.S. Forest Service provides examples in nature where temperature changes can affect ecosystems and the entire biome. The key lesson here is that the biological organisms are usually the smaller players in the larger drama between the land masses and the patterns of weather developing in each biome.

As you watch the program, consider the following questions:

1. Is the Clapper Rail headed for extinction or is the species having a temporary decline due to excessive predation, lack of resources, a disruption in the lower trophic levels, or some other reason? Explain your response as it relates to the specific ecosystem.

2. If nature had previously recycled nutrients using forest fires as the mechanism, why must we use "prescribed burns" to accomplish the same results today?

3. What are the similarities and differences between the function of the decomposers and the effects of periodic fires?

4. Can energy be recycled?

5. Why do ecological pyramids have a wide base and a narrow tip?

6. What are some of the methods used to study weather patterns? Do you think the data collected can help to predict future climatic conditions?

7. Are temperature gradients present in all biomes? What is one example of an adaptation to a colder temperature range?

8. Can you also give an example of adaptation to a very arid environment?

Review Activities

Matching

Match the terms listed below with the definitions that follow. Check your answers with the Answer Key and review any terms you missed.

I.

____ 1. ecosystem modeling   ____ 6. detrital food web
____ 2. food web   ____ 7. consumer
____ 3. trophic levels   ____ 8. decomposer
____ 4. food chain   ____ 9. grazing food web
____ 5. primary producers   ____ 10. ecological pyramids

a. a network of food chains cross-connected among primary producers, consumers, decomposers, and detritivores

b. a flow of energy from plants to decomposers and detritivores

c. a sequence of consumption as one organism is eaten by a second one, which in turn is eaten by a third, and so on

d. predictions made using computers and models to determine possible effects of disturbances within the ecosystem

e. a flow of energy from plants to herbivores and then to carnivores

f. organisms transferring energy and other resources by feeding on tissues of other organisms

g. levels defined by a hierarchy of feeding and the same number of steps from the original energy source

h. concept of ecosystems as primary producers at the base with various trophic levels of consumers above them

i. organisms that break down wastes, by-products, and dead remains of other organisms to obtain necessary energy

j. plants that convert and capture energy from the sun and other resources

II.

___ 1. energy pyramids
___ 2. biogeochemical cycle
___ 3. water cycle
___ 4. watershed
___ 5. nitrogen cycle
___ 6. carbon cycle
___ 7. phosphorus cycle
___ 8. greenhouse effect
___ 9. biological magnification
___ 10. leaching

a. movement of carbon-based molecules from reservoirs such as the atmosphere and the ocean through organic life-forms in various ecosystems and back again

b. movement of a variety of elements and compounds from their environmental sources to living organisms and back to the environment

c. sedimentary cycling of phosphorus from land to sea and back to land with a rapid loop through the ecosystems from plants into herbivores to higher trophic levels and back again

d. a hydrologic cycle driven by solar energy, weather patterns, topography, and aquatic ecosystems

e. an increase in effect or concentration of an element moving from lower to higher trophic levels or along the food chain

f. movement of certain nutrients or elements out of one area into another by physical mechanisms

g. the interchange and movement of nitrogen-based molecules from the atmosphere to the oceans, land masses, and through the ecosystems and back again

h. concept of the trophic levels in an ecosystem as represented by the content or loss of energy at each level

i. warming of the biosphere due to the effects of certain gases such as carbon dioxide and nitrous oxide

j. a regional source of water flow created when precipitation is concentrated into one river or suitable conduit

III.

___ 1. rain shadow
___ 2. biome
___ 3. climate
___ 4. realms
___ 5. estuaries
___ 6. tropical rain forests
___ 7. deserts
___ 8. prairies
___ 9. forests
___ 10. El Niño

a. major land regions defined by geographic and climatic factors which have the effect of isolating various ecosystems

b. a region of reduced precipitation caused by topography or weather patterns

c. a biome defined by arid conditions and reduced ground cover exhibiting ecosystems having unique adaptations to heat, sparse vegetation and lack of extensive watersheds

d. a biome defined by ecosystems containing tall canopies of trees and several habitats from the ground upward

e. a biome defined by heavy precipitation, canopies of tall trees, mean temperatures of 25 degrees centigrade, and ecosystems containing great biodiversity

f. a section of a realm defined by certain characteristic types of vegetation, climates, topography, and soils

g. an effect of changes in surface ocean currents originating in the Pacific Ocean, capable of contributing to worldwide changes in a variety of biomes

h. a biome defined by grasslands having a flat or gently rolling topography with thickly rooted topsoil

i. a long-term pattern of weather conditions affected by the proximity of water and land masses

j. coastal regions where freshwater flows into a partly enclosed land mass allowing seawater to encroach and produce unique ecosystems

Completion

Fill in each blank with the most appropriate term from the list for that paragraph. A term may be used once, more than once, or not at all. If a question requires two or more answers in succession, they may be in any order (unless the question indicates otherwise). Check your answers with the Answer Key and review when necessary.

1. A model of an ecosystem can help to predict effects of _____ to the system. The basic structure of the model shows that _____ flows in only one direction while various _____ are cycled within the _____. Primary producers, also known as _____, can store and convert some energy from the _____. Nutrients are transferred to organisms known as _____ which in turn _____ the nutrients back to the autotrophs. Three types of heterotrophs are _____, _____, and _____. While nutrients are recycled, the energy is lost during the process and returns to the environment. Thus there is a constant need for new energy no matter how efficient the ecosystem has become. Using the various _____ and outputs of the model, _____ simulations can help study these complex interactions. It is important to note that the _____ success of the _____ will depend upon the accuracy of the parameters included.

autotrophs	hydrosphere
biosphere	inflows
computer	inputs
consumers	model
decomposers	moon
detritivores	nutrients
disturbances	predictive
ecosystem	producers
energy	recycle
heterotrophs	sun

322   Lesson 25 / Ecosystems and the Biosphere

2. An example of a _____ cycle is the nitrogen cycle. The cycle can begin with nitrogen _____ where bacteria converts atmospheric _____ to ammonia and ammonium. The _____ use this converted nitrogen which becomes the only source for _____. Certain other bacteria and _____ can break down the wastes and dead tissues into usable nitrogen compounds. Still other bacteria convert these compounds using _____ into _____ and further into _____ which is once again taken up by the plants. In another process known as _____, atmospheric nitrogen is produced, escaping the ecosystem. Finally, the human factor of the nitrogen cycle includes the use of _____ and the burning of _____ fuels that change the variables in the nitrogen cycle.

animals	nitrification
biogeochemical	nitrite
chlorophyll	nitrogen
denitrification	oxygen
fertilizers	plants
fixation	sunlight
fossil	wax
fungi	release
nitrate	

3. The biosphere is more than the sum of its parts. The interrelationships between the ocean, the land, and the _____ are much more complex. A good example of this is the El Niño phenomenon and its far-reaching effects. For reasons not well understood, a _____ current surfaces in the South Pacific Ocean, flowing _____ toward the coast of _____. The colder _____ current is displaced and with increased temperatures and _____, the air pressure drops. Changes then occur in the weather patterns resulting in destructive storms along the _____, heavier rainfall inland and lack of _____ in other areas. Agricultural effects include _____ of crops or reduced _____ due to variations in rainfall and even _____. What began as a wave of warm water ended up having _____ effects on many ecosystems and _____, including our own.

coasts	Peru
mountains	Canada
east	warm
west	loss
atmosphere	drought
lithosphere	evaporation
Sargasso	precipitation
Humboldt	populations
productivity	global

Self-Test

Select the one best answer for each question. Check your answers with the Answer Key and review when necessary.

1. Most of the energy in an ecosystem is lost due to
   a. degeneration of soil.
   b. recycling of nutrients.
   c. metabolic heat.
   d. the nitrogen cycle.

2. An example of an organism at the second trophic level is
   a. a tiger.
   b. a redwood tree.
   c. a cow.
   d. a fish tapeworm.

3. Ecological pyramids are characterized by
   a. energy inputs at each trophic level.
   b. biomass distribution at each trophic level.
   c. efficiency of nutrient recycling.
   d. geographic location.

4. Biogeochemical cycles are important in the process of
   a. temperature stabilization.
   b. introduction of new species into the ecosystem.
   c. recycling of nutrients.
   d. all of the above.

5. All of the following are examples of a biogeochemical cycle EXCEPT
   a. metamorphic cycles.
   b. atmospheric cycles.
   c. sedimentary cycles.
   d. hydrologic cycles.

6. One example of biological magnification is
   a. radioactive fallout in the grazing pastures of milk cows.
   b. the El Niño effect on the Humboldt current.
   c. extreme high tides.
   d. meteor showers in desert biomes.

7. Global air circulation patterns and climatic conditions are affected by
   a. the rotation of the Earth.
   b. variations in landmass distribution.
   c. latitude differences.
   d. all of the above.

8. A biome is defined by all the following EXCEPT
   a. characteristic vegetation.
   b. soil types.
   c. part of a realm.
   d. number of trophic levels.

9. All of the following affect zonation in aquatic ecosystems EXCEPT
   a. depth gradients at the shore level.
   b. primary forest succession.
   c. amount of sunlight daily.
   d. range of rooted aquatic plants.

10. Which of the methods below are used to study ecosystems?
    a. weather satellites
    b. tracking consumer populations
    c. ecological modeling
    d. all of the above

Using What You've Learned

Based on your own interests or your instructor's requirements, complete one or more of the following activities.

1. Construct a diagram of the food web to which you belong. Remember to include all cross-connections, waste, and interrelationships with other organisms.

2. Using charts, drawings, and/or photographs, describe your biogeographic realm, biome, and ecosystem to which you belong. Be as specific as possible.

3. Choose one of the major biomes discussed in the text and investigate the disturbances attributable to human intervention. Discuss how some of the ecosystems in the biome might be returned to their more natural state.

4. In terms of energy flows, describe the most efficient human diet that will also conserve as many resources as possible in order to feed a hungry world.

5. Visit your local library and research the 1991 treaty banning mineral exploration in Antarctica for the next fifty years. What additional provisions would you include to further protect and preserve the continent?

6. Consider the evidence for global warming as presented in the text. Could this be simply another cycle which repeats itself periodically or is this a unique situation? Support your answer after completing additional research into long-term patterns of climatic changes over the past several million years. How much warming or cooling might we expect over the long run?

Challenge Questions

1. Certain primary producers can also function as primary and secondary carnivores. Examples include the Venus flytrap, the pitcher plant, and the sundew. In terms of energy needs and ecosystems, can you explain these dual roles?

2. What is the difference when a watershed is lost due to sudden deforestation versus a gradual succession to a new vegetation?

3. Other than human intervention, what are several events that could lead to eutrophication in aquatic ecosystems?

4. Do you consider the moon to be a part of the human ecosystem? Is the ocean floor where Sealab is located part of the human ecosystem?

5. Assume Mt. Lassen in California has just erupted after a period of dormancy. If scientists determine that airborne particulates are one hundred times greater than those measured by the recent eruption of Mt. St. Helens, what are some effects we might expect to see both regionally and globally?

LESSON 26

The Human Factor

Assignments

For the most effective study of this lesson, we suggest that you complete the assignments in the sequence listed below:

Before Viewing the Video Program

- Read the Overview and Learning Objectives for this lesson. Use the Learning Objectives to guide your reading, viewing, and thinking.

- Read Chapter 43, "Human Impact on the Biosphere," pages 752–769, in the Starr textbook.

  Or read, in the Starr/Taggart textbook, Chapter 50, "Human Impact on the Biosphere," pages 882–899.

- Read the Viewing Notes in this lesson.

View "The Human Factor" Video Program

After Viewing the Video Program

- Briefly note your answers to the questions at the end of the Viewing Notes.

- Review all reading assignments for this lesson, especially the Chapter 43 Summary on page 768 in the Starr textbook (Chapter 50 Summary on page 908 in the Starr/Taggart textbook) and the Viewing Notes in this lesson.

- Write brief answers to all the Review Questions at the end of Chapter 43 in the Starr textbook (Chapter 50 in the Starr/Taggart textbook) to be certain you understand the text material.

- Complete the Review Activities in this guide to reinforce your understanding of important terms and concepts. Check your answers with the Answer Key and review when necessary.

- Take the Self-Test in this guide to measure your achievement of the Learning Objectives. Check your answers with the Answer Key and review when necessary.

- Complete the Using What You've Learned activities and any other activities and projects assigned by your instructor.

Overview

We began our study of ecology by exploring the concepts that define various populations and communities. To fully understand the relationships between plants, animals and their unique environments, three essential questions must be answered: "Where do the organisms live? How do they live there? And why do they live there?" When applied to communities, the answers provide clues to the biodiversity of the living world. A more global picture emerges as the same questions are asked about ecosystems, biomes, and the entire biosphere. One important answer to these questions can be summarized as follows: When the land, oceans, and atmosphere interact with each other, the effects are felt throughout the biosphere and define the conditions under which life must adapt.

There is of course, another variable which looms large in this equation: the human factor. Lesson 26 will address our role in the biosphere, focusing upon the changes humans have made and continue to make today. First, the impact of air pollution and acid rain presents a picture of a vulnerable biosphere. Expanding on this theme is a discussion of the effects of ozone depletion, waste accumulation and limitations to agricultural expansion. While the long-term effects of these activities are being debated, you will see how the destruction of the world's great forests is now making irreversible changes in the major biomes.

The lesson continues with a closer look at the effects of agriculture on the land and the water needed to sustain us all. As you will see, while agricultural success is being achieved in the "green revolution," growth of the human population continues to add stresses to land use, ecosystem sensitivity and the world's resources in general. Of course, a major limiting resource is the availability of usable water. Also driving the entire human factor is the never-ending consumption of energy. Recall from your study of basic metabolism that loss of energy through chemical reactions and heat will only intensify the search for new and greater energy sources. Thus while the laws of thermodynamics may seem like only an academic concept, there is no escaping their eventual reality. Certainly, technology may provide future sources of energy but questions concerning pollution and by-product disposal remain to be answered.

Finally, a dilemma emerges for the current and future generations of human beings. This dilemma can be stated as follows: Given that the biosphere is governed by rules that change very little over time, how can we preserve the biosphere and at the same time provide for the well-being of the human population? As you complete your study of the *Cycles of Life,* it is hoped that you will address this and other questions that affect your life and the world we live in.

Learning Objectives

When you have completed all assignments for this lesson, you should be able to:

1. Relate the first and second laws of thermodynamics to the use of world resources and food supplies, and discuss the potential for all people on Earth to enjoy the benefits of an industrialized lifestyle.

2. Discuss ways humans are and are not unique among organisms in their relationship to the environment, and generally discuss the impact of humans on the ability of ecosystems to be self-sustaining.

3. Identify the major classes of air pollutants and their sources, and discuss the effect of these pollutants on the Earth's biosphere.

4. Identify the major sources of water pollution, describe their general effects on living organisms, and discuss the challenges of maintaining water quality in the modern world.

5. Examine the effects of human land use decisions on various ecosystems of the Earth, noting the roles of agriculture, *desertification*, and *deforestation*.

6. Use the human growth curve and the need for increased agricultural productivity to show the interrelationship of changes in water use, land usability, atmosphere, ecosystem diversity, and energy consumption.

7. Distinguish between *renewable* and *nonrenewable* energy sources. Discuss the current level of use for each and the promise they hold for the future.

8. Describe how scientists study ecological changes, both on a small scale and in the biosphere as a whole.

Viewing Notes

The final video for this course, "The Human Factor," is a study in human behavior. It is a story of environmental successes combined with failures, of unwitting mistakes and their long-term effects, and of the lessons we must learn in order to improve our efforts in dealing with the natural world. The story centers on the inescapable fact that life is interdependent, that we all depend on the environment for our most basic needs: air to breathe, water to drink, food to eat, space in which to live. Of all living things on this planet, humans are unique in their global impact on other organisms and on the life-sustaining environment itself. Yet humans are also unique in their ability to make measured choices, devise creative solutions, and predict consequences. In that ability lies hope for the future of life on Earth. This program integrates accounts of how we can use our scientific knowledge to solve the problems created by the spread of industrialization with views of the positive role biology may play in shaping the environment of the twenty-first century.

Biosphere Earth is a complex and fragile place. In this video you'll learn about two issues that have dominated the headlines over the last decade and are still at the forefront of global environmental concerns: ozone layer reduction and global warming. One segment examines the importance of predictive models for the factors at work in both issues. Among the examples of modeling is the work of Dr. Walter Oechel, San Diego State University, in whose field experiments you'll observe carbon dioxide accumulation and learn about the effect of "greenhouse gases" on global warming and of potential climatic changes on ecosystems.

A video segment addresses how our understanding of the biological and physical world will help lessen the human impact on fragile ecosystems—in the resources we take, the wastes we leave, the land we use, and the food and other resources we need to survive. It is these issues that concern Dr. William Frankenberger, University of California, Riverside, as he strives to create a microbiological answer for neutralizing toxins (both natural and those created by humans) in the soil and in runoff water from agricultural lands. Working in conjunction with the USDA, Dr. Frankenberger developed a method to enlist naturally occurring microbes in the task of detoxifying areas of selenium-tainted runoff in California's San Joaquin Valley. He devised a process of soil amendment called microbial volatilization, which speeds up the process of natural detoxification. He demonstrated that if humans help nature, rather than overwhelm it, natural processes can help overcome some forms of environmental pollution.

Another segment shows how the search for clean energy sources is providing new hope for better tomorrows. The video shows a hydrogen-powered car, which produces virtually no pollution, and considers the possible contributions of fusion reactors, which offer seemingly limitless power with none of the radiation problems of atomic fission power plants. Total mastery of our energy situation can be achieved, however, only if we place conservation, efficient energy use, and a stable population that ensures predictable energy needs among our highest priorities. This video will aid your comprehension and appreciation of these matters by presenting Dr. Susan Hackwood, University of California, Riverside, current Executive Director of the California Council on Science and Technology, who comments insightfully on conservation ecology.

As you watch the program, consider the following questions:

1. Why do human beings play a unique role both in the degradation and the preservation of the biosphere?
2. What is the importance of the "greenhouse effect"?
3. Could global warming be simply a part of the long-term climatic cycles which include periodic ice ages as well?
4. Why can't microbes that naturally clean up the environment continue to do so as in the past? How have humans made their task harder? Easier?
5. Why is energy from fusion reactors, unlike that from fission reactors, so attractive from an environmental perspective?
6. Are all attempts to recreate an ecosystem doomed to failure?
7. How far should we go to protect and nurture endangered species?
8. Are the effects of pollution reversible?
9. Do you think humans can change the behaviors leading to disruptions in the environment?
10. What is meant by the idea that all life on this planet is interdependent?

Review Activities

Matching

Match the terms listed below with the definitions that follow. Check your answers with the Answer Key and review any terms you missed.

I.

___ 1. industrial smog
___ 2. photochemical smog
___ 3. thermal inversion
___ 4. PANs
___ 5. ozone
___ 6. nitric oxide
___ 7. chlorofluorocarbons
___ 8. sulfur dioxide
___ 9. hydrocarbons
___ 10. acid rain

a. reacts with ozone to produce oxygen

b. nitrogen-containing pollutant precursor to photochemical oxidants

c. gray haze pollution from fossil fuels and coal

d. warm air layer trapping polluted cold air

e. pollutant converts to sulfuric acid in water

f. brown haze pollution containing oxidants

g. petroleum distillates and other H-C compounds

h. oxidants containing peroxyacyl nitrates

i. atmospheric pollutants converted to acids

j. oxidants in smog and protective atmospheric layer

II.

____ 1. deforestation   ____ 6. water table
____ 2. shifting cultivation ____ 7. Chernobyl
____ 3. desertification   ____ 8. *Exxon Valdez*
____ 4. desalination   ____ 9. fusion
____ 5. Ogallala aquifer   ____ 10. solar-hydrogen energy

a. removal of salts from water

b. source of large oil spill in Alaska

c. groundwater source of 20% of U.S. irrigation needs

d. sunlight aids splitting of water into O_2 and H_2 gases

e. short-sighted technique for creating farmland

f. atomic nuclei fuse, producing energy

g. massive removal of trees from indigenous regions

h. source of radioactive meltdown in Russia

i. grasslands and croplands converted to desert

j. maximum expected level of groundwater

III.

____ 1. subsistence agriculture
____ 2. animal-assisted agriculture
____ 3. mechanized agriculture
____ 4. high-yield crops
____ 5. greenhouse effect
____ 6. alkaloids
____ 7. thermal pollution
____ 8. nuclear reactors
____ 9. wind farms
____10. feedforward mechanisms

a. cultivation aided by oxen and horses among others

b. extracts from tropical rain forest plants used to treat heart problems and cancer

c. machines used for planting, fertilizing, pesticides, watering, and harvesting

d. convert heat from nuclear rods to electricity

e. labor-intensive cultivation for immediate needs only

f. efficient and bountiful species of food-producing plants

g. Factories and other local causes of temperature changes enhancing disruption in the surrounding environment

h. early warning of a larger problem in the future

i. groups of wind-collectors capable of producing electricity

j. gases and massive burning causing a warming of the lower atmosphere

Completion

Fill in each blank with the most appropriate term from the list for that paragraph. A term may be used once, more than once, or not at all. If a question requires two or more answers in succession, they may be in any order (unless the question indicates otherwise). Check your answers with the Answer Key and review when necessary.

1. The forests of the world provide habitats for a majority of land-based life forms. Unfortunately, deforestation threatens this rich resource. This practice is proceeding due to the need for new _____, _____ land for animals and demands for _____ and _____ to build shelters. Deforestation occurs in tropical as well as _____ forests. Currently the areas of greatest destruction are _____ and _____ in South America, Mexico, and Indonesia. Effects of the shrinking forests include soil _____, _____ changes, loss of _____, _____ of rivers, plant and animal _____, and amplification of the _____ effect. Unless drastic measures are taken to stop this destruction, entire biomes may be converted to _____ or even new deserts.

    Australia   grasslands
    biodiversity   greenhouse
    Brazil   growth
    breeding   landmass
    Colombia   lumber
    cropland   more grazing
    erosion   temperate forests
    extinctions   tundra
    flooding   weather
    fuel

2. Using alternate forms of energy may reduce several _____ impacts on the environment. _____ energy can tap the sun's rays and create electricity directly from _____ panels, by heat conversion and in combination with other technologies. One combination uses the _____ from the panels to separate water into hydrogen gas and oxygen. This _____ energy uses clean-burning H₂ gas, eliminating _____ seen in fossil fuels. Another source of energy is wind power. Wind _____ capture this energy which can be added to the power grid or stored in _____. Some other sources may be used selectively including _____ conversion, tapping _____ outlets including volcanoes, and even expanding _____ power to the tidal zones. Perhaps the most enticing new resource is the safe utilization of fusion. True _____ would provide for a virtually _____ source of energy, but researchers have yet to harness the reactions involved. While looking for the fusion answer, other energy sources such as hydrogen _____ driven by _____ power in magnetic fields show additional promise.

batteries	hydroelectric
biomass	implosion
climatic	laser
electricity	nuclear
emissions	photovoltaic
explosion	solar
fission	solar-hydrogen
fusion	turbines
geothermal	unlimited
heat	

3. Water is a crucial factor in the long-term survival of the _____ as we know it. Since _____ is not a cost-effective solution, conservation of _____ resources is imperative. Contamination of this resource can be reduced by taking toxic substances out of the environment, recycling others and using effective _____ treatments. The _____ treatment uses gravity tanks and filtering techniques to separate sludge. Pathogens are controlled by _____ but residual toxins may remain. _____ treatments involve _____ "scrubbers" to clear other wastes and toxins. The water is again treated with chlorine prior to release. _____ treatment uses advanced _____ to absorb toxins, convert substances to harmless compounds, and eliminate pathogens and _____. It is effective but _____. Since the _____ of water affects us all, each nation must be responsible for maintaining usable water supplies. Only a coordinated effort will achieve this goal.

antibiotics	primary
biosphere	quality
chlorination	quantity
desalination	saltwater
evaporation	secondary
expensive	technology
freshwater	tertiary
microbial	thermal
pollutants	wastewater

Self-Test

Select the one best answer for each question. Check your answers with the Answer Key and review when necessary.

1. Global industrialization is unlikely to occur due to
   a. temperature extremes in the northern latitudes.
   b. massive energy needs.
   c. the greenhouse effect.
   d. none of the above.

2. Human engineered ecosystems may be limited by
   a. surface area.
   b. their ability to be self-sustaining.
   c. production of antioxidants.
   d. their loss of methyl bromide.

3. Production of photochemical smog is initially triggered by
   a. sulfur oxides.
   b. polyaromatic hydrocarbons.
   c. nitric oxide.
   d. all of the above.

4. Dry acid deposition can be converted to acid rain when
   a. temperatures are above 30 degrees centigrade.
   b. oxides dissolve in atmospheric water.
   c. the pH of soil is above 9.5 units.
   d. a thermal inversion occurs.

5. Major sources of water pollution include
   a. pesticides.
   b. poorly treated wastewater.
   c. discharges by factories.
   d. all of the above.

6. All of the following are examples of poor land use decisions **EXCEPT**
    a. designated wilderness areas.
    b. desertification.
    c. deforestation.
    d. nuclear waste disposal.

7. The human growth curve matches
    a. decreases in energy consumption.
    b. decreases in water consumption.
    c. increases in the ozone layer.
    d. increases in land cultivation.

8. All of the following energy sources are renewable **EXCEPT**
    a. solar.
    b. wind.
    c. fossil fuels.
    d. solar-hydrogen.

9. Human beings' impact on the global biosphere
    a. is mostly positive.
    b. is somewhat negative.
    c. is entirely negative.
    d. can be both positive and negative.

10. Potential problems for the biosphere are best solved by
    a. feedback control.
    b. feedforward mechanisms.
    c. anticipation of future events.
    d. technological advances.

Using What You've Learned

Based on your own interests or your instructor's requirements, complete one or more of the following activities.

1. Obtain a copy of the water quality report from your local supplier. This is a public record and is sent to each customer annually. Are you surprised at the number and types of compounds listed?

2. Going one step further, compare these compounds with a current list of known or suspected carcinogens. Now consider the following question: While all the compounds, including the carcinogens, are below the established levels, what are the implications for long-term consumption of this water?

3. Describe your local city- or county-sponsored recycling program. What additional materials would you like to see recycled? Can you recycle these privately?

4. Make a list of mainstream environmental organizations and another list of more activist groups. Compare and contrast the effectiveness of the two approaches.

5. Prepare a speech for a high school audience on Earth day. What would be your main theme and how would you inspire the students to action?

6. Much time, effort, and money have been spent to correlate pollution levels with the prevalence of certain diseases. Studies have also matched specific areas to the risks of contracting those diseases. Should a current summary of this research be available to the public? Can you find any of these studies in your local library?

7. Devise an ecosystem that will be self-sustaining and place it within the city limits where your home is located. Describe the environmental pressures on the ecosystem and speculate on the long-term effects you would expect to see.

8. Locate an example of an environmental impact report. Using similar criteria and your knowledge from this biology course, outline a report on a parcel that is zoned for industrial use.

Challenge Questions

1. What incentives could you give to industrialized and developing countries in order to cooperate in finding solutions to environmental problems?

2. Assume you are the last human being to have lived in a tropical rain forest before they were completely destroyed. Further assume the technology exists to establish a new rain forest on the planet. As a consultant to the project, you are asked to choose the location and the parameters to insure its success. What recommendations would you make?

3. If you could choose only one alternative energy source to develop, which one would it be? Justify your answer.

4. Change has always been a part of the biosphere. Are we overreacting or underreacting to the human contribution to these changes?

5. Using the knowledge gained from this course, describe the adaptations of structure and function that animals and plants might acquire in response to the conditions caused by pollution over a long period of time.

6. Assume Earth were in the path of a huge comet and the collision could not be avoided. If a single, large spacecraft were available for escape, what life-forms would you take on this new Noah's ark?

7. How would you preserve the remaining wilderness areas on Earth? Would these areas be actively managed or simply left alone?

Answer Key

Lesson 1

Matching

I.

1. c
2. a
3. e
4. f
5. d
6. b

II.

1. g
2. f
3. e
4. b
5. h
6. c
7. a
8. d

III.

1. e
2. d
3. g
4. h
5. b
6. i
7. a
8. f
9. c

Completion

1. unity, diversity, substances, physical, metabolism, environment, DNA, diversity, environment, traits, environment
2. evolution, natural selection, kingdoms, species, population, trait, traits, survive, reproduce, traits, traits, population
3. scientific inquiry, problem, question, hypothesis, question, hypothesis, predictions, experiments, experiments, observations, observations, experiments, conclusions, hypothesis

Self-Test

(S after page numbers refers to the Starr textbook, S/T after page numbers refers to the Starr/Taggart textbook)

1. c (Objective 1; page 4 S; page 4 S/T)
2. d (Objective 2; page 7 S; page 7 S/T; video program)
3. d (Objective 3; page 10 S; page 10 S/T)
4. a (Objective 4; page 8 S; page 8 S/T [with Monera described as comprising Archaebacteria and Eubacteria; see also page 329])
5. a (Objective 5; pages 10–11 S; pages 10–11 S/T; video program)

Answer Key 343

6. b (Objective 6; video program)
7. c (Objective 7; video program)
8. a (Objective 8; page 12 S; page 12 S/T; video program)
9. d (Objective 9; page 13 S; page 13 S/T; video program)
10. a (Objective 10; page 15 S; page 16 S/T; video program)

Lesson 2

Matching

I.

1. d
2. i
3. k
4. e
5. j
6. h
7. a
8. c
9. b
10. f
11. g
12. l

II.

1. i
2. f
3. b
4. g
5. d
6. h
7. j
8. e
9. c
10. a

Completion

1. elements, element, atoms, protons, neutrons, electrons, protons, neutrons, electrons, orbitals, ion, isotopes

2. atoms, molecule, molecules, compounds, electron, ionic bond, electrons, covalent bond, molecule, polar bond, polar bond, solvent, release, combine with

3. carbon, organic compounds, organic compounds, proteins, carbohydrates, lipids, nucleic acids, carbohydrates, lipids, proteins, nucleic acids

Self-Test

(S after page numbers refers to the Starr textbook, S/T after page numbers refers to the Starr/Taggart textbook)

1. c (Objective 1; page 22 S; page 22 S/T; video program)
2. c (Objective 2; page 22 S; page 22 S/T; video program)
3. b (Objective 3; page 22 S; pages 22–23 S/T; video program)
4. d (Objective 4; video program)
5. b (Objective 5; pages 24–25 S; pages 26–27 S/T; video program)
6. d (Objective 6; pages 26–27 S; pages 28–29 S/T; video program)
7. b (Objective 7; page 29 S; page 30 S/T; video program)
8. b (Objective 8; page 36 S; page 38 S/T; video program)
9. d (Objective 9; page 37 S; page 40 S/T; video program)
10. c (Objective 10; pages 42–43 S; page 46 S/T; video program)

Lesson 3

Matching

I.

1. g	6. b	11. f
2. n	7. l	12. i
3. e	8. k	13. m
4. c	9. d	14. a
5. h	10. j	

II.

1. g	6. l	11. e
2. f	7. i	12. b
3. m	8. j	13. c
4. h	9. k	14. n
5. a	10. n	15. d

III.

1. d	4. a	7. b
2. c	5. g	8. e
3. f	6. h	

Completion

1. compound microscope, lenses, cells, cell theory, all, smallest, life, growth, division, single

2. fluid mosaic model, membrane, lipid bilayer, phospholipids, proteins, hypertonic, hypotonic, diffusion, transport proteins

3. plasma, eukaryotic cells, cytoplasm, organelles, nucleus, DNA, mitochondria, ribosomes, endoplasmic reticulum, proteins, prokaryotic cells, organelles, cytoskeleton, microtubules, microfilaments, flagella

4. plant, cell walls, chloroplasts, sunlight, chemical bonds, water, carbon dioxide, central vacuole

Self-Test

(S after page numbers refers to the Starr textbook, S/T after page numbers refers to the Starr/Taggart textbook)

1. b (Objective 1; page 51 S; page 56 S/T; video program)
2. d (Objective 2; pages 52–53 S; pages 56–57, 82–83 S/T; video program)
3. c (Objective 3; page 84 S; page 86 S/T)
4. b (Objective 4; pages 51–52, 84–85 S; pages 56–57, 86–87 S/T)
5. a (Objective 5; pages 52, 84–85 S; pages 56–57, 86–89 S/T)
6. c (Objective 6; page 53 S; pages 82–83 S/T; video program)
7. d (Objective 7; page 70 S; page 56 S/T; video program)
8. a (Objective 8; pages 56, 60–61 S; page 64 S/T; video program)
9. b (Objective 9; page 65 S; page 69 S/T)
10. d (Objective 10; page 56 S; pages 60–63 S/T; video program)

Lesson 4

Matching

I.

1. f
2. d
3. a
4. h
5. b
6. e
7. c
8. g

II.

1. b
2. d
3. k
4. g
5. h
6. i
7. f
8. i
9. c
10. e
11. a

Completion

1. first law of thermodynamics, energy, second law of thermodynamics, increases, entropy, sunlight, chemical bonds, metabolism

2. reactants, products, reversible, reversible, acid, base, reversible, reactants, products, equilibrium, endergonic, exergonic, exergonic, endergonic, coupled reaction

3. metabolic pathways, ATP/ADP cycle, ATP, phosphate group, ATP, released, phosphorylation

4. enzymes, speed up, substrate, activation energy, induced-fit model, substrate, enzyme, active site, temperature, pH, cofactors, coenzymes, electrons, electron transport system

Self-Test

(S after page numbers refers to the Starr textbook, S/T after page numbers refers to the Starr/Taggart textbook)

1. c (Objective 1; page 76 S; page 98 S/T; video program)
2. a (Objective 2; pages 76–77 S; pages 98–99 S/T; video program)
3. d (Objective 3; page 80 S; pages 100–101 S/T; video program)
4. a (Objective 4; pages 80–81 S; page 104 S/T; video program)
5. b (Objective 5; page 75 S; page 6 S/T)
6. d (Objective 6; page 79 S; page 107 S/T; video program)
7. d (Objective 7; pages 78–79 S; page 101 S/T; video program)
8. b (Objective 8; page 79 S; page 102 S/T; video program)
9. a (Objective 9; pages 74–75 S; page 109 S/T; video program)

Lesson 5

Matching

I.

1. b	6. c	10. i
2. e	7. k	11. d
3. h	8. g	12. f
4. l	9. a	13. j
5. m		

II.

1. d	5. a	8. i
2. c	6. j	9. f
3. h	7. b	10. e
4. g		

Completion

1. carbon, energy, light-dependent, chloroplasts, energy, chemical bond, ATP, ATP, NADPH, light-independent, stroma, sucrose, starch
2. phosphate, ATP, aerobic respiration, glycolysis, glucose, two, two, four, Krebs cycle, one, three, electron transport phosphorylation, mitochondria, 36, glucose

Self-Test

(S after page numbers refers to the Starr textbook, S/T after page numbers refers to the Starr/Taggart textbook)

1. a (Objective 1; page 92 S; page 112 S/T; video program)
2. a (Objective 2; page 92 S; pages 114–115 S/T; video program)
3. b (Objective 2; pages 94–95 S; pages 114–115 S/T; video program)
4. b (Objective 3; video program)
5. c (Objective 4; video program)
6. c (Objective 5; page 110 S; page 132 S/T)
7. d (Objective 6; page 118 S; page 140 S/T)
8. d (Objective 7; pages 120–121 S; page 142 S/T)
9. a (Objective 8; page 122 S; page 144 S/T)
10. b (Objective 9; video program)

Lesson 6

Matching

I.

1. c
2. d
3. g
4. i
5. b
6. j
7. a
8. e
9. h
10. f

II.

1. f
2. j
3. d
4. a
5. c
6. i
7. h
8. g
9. e
10. b

Completion

1. reproduction, prokaryotic, binary fission, eukaryotic, mitosis, meiosis, mitosis, the same, meiosis, halves, gamete, half, gamete

2. sexual reproduction, asexual reproduction, sexual reproduction, [meiosis, gamete formation, fertilization], meiosis, diploid, germ cells, meiosis, cross over, haploid, gametes, haploid, gametes, fertilization

3. cell cycle, interphase, cell cycle, mitosis, chromosomes, condense, break down, prophase, metaphase, breaks down, chromosomes, spindle apparatus, anaphase, sister chromatids, telophase

Self-Test

(S after page numbers refers to the Starr textbook, S/T after page numbers refers to the Starr/Taggart textbook)

1. d (Objective 1; page 128 S; page 150 S/T; video program)
2. c (Objective 2; page 129 S; page 151 S/T; video program)
3. b (Objective 3; pages 130–131 S; pages 152–153 S/T)
4. b (Objective 4; page 133 S; page 157 S/T)
5. c (Objective 5; page 135 S; page 158 S/T)
6. a (Objective 6; page 140 S; page 162 S/T; video program)
7. d (Objective 7; page 140–141 S; pages 162–163 S/T; video program)
8. a (Objective 8; pages 142–143 S; pages 166–167 S/T)
9. a (Objective 9; page 149 S; pages 170–171 S/T; video program)
10. c (Objective 10; video program)

Lesson 7

Matching

I.

1. h
2. b
3. f
4. d
5. j
6. a
7. g
8. c
9. i
10. e

II.

1. a
2. c
3. b
4. d
5. e
6. h
7. g
8. i
9. f

III.

1. b
2. j
3. f
4. c
5. g
6. a
7. d
8. e
9. i
10. h

Completion

1. monohybrid, diploid, genes, genes, dominant, inheritance, segregation, gametes, independent assortment

2. Punnett squares, gene interactions, environmental factors, incomplete dominance, codominance, pleiotropy, epistasis

3. sex-linked, hemophilia, recessive, X-linked, crossing over, crossing over, linked, meiosis, nondisjunction, Down syndrome, Turner syndrome, Klinefelter syndrome, deletion, inversion, translocation, duplication

Self-Test

(S after page numbers refers to the Starr textbook, S/T after page numbers refers to the Starr/Taggart textbook)

1. b (Objective 1; page 156 S; pages 178–179 S/T; video program)
2. c (Objective 1; page 159 S; pages 158–159 S/T; video program)
3. a (Objective 2; pages 156–157 S; pages 156–157 S/T; video program)
4. b (Objective 3; page 160 S; page 182 S/T; video program)
5. d (Objective 4; page 176 S; pages 198–199 S/T; video program)
6. d (Objective 5; page 172 S; page 194 S/T; video program)
7. c (Objective 6; pages 184–185 S; pages 208–209 S/T; video program)
8. d (Objective 6; pages 184–185 S; pages 208–209 S/T; video program)

9. b   (Objective 7; page 178 S; page 202 S/T; video program)
10. d   (Objective 8; page 186 S; page 212 S/T; video program)

Lesson 8

Matching

I.

1. e
2. f
3. c
4. b
5. d
6. a
7. g
8. h

II.

1. d
2. c
3. f
4. a
5. e
6. g
7. b

Completion

1. DNA, Miescher, Griffith, bacterial, Hershey, Chase, radioactive isotopes, DNA, X-ray crystallography, Watson, Crick, double-stranded helix

2. nucleotides, five, phosphate, adenine, guanine, thymine, cytosine, A-T, C-G, chromosome, hydrogen, enzymes, polymerases, nucleotides, ligases, nucleotide

3. genetic engineering, recombinant DNA technology, DNA fragments, host cells, recombinant DNA technology, restriction enzymes, DNA strands, DNA fragments, plasmids, cloning, vectors, viruses, host cells

Self-Test

(S after page numbers refers to the Starr textbook, S/T after page numbers refers to the Starr/Taggart textbook)

1. d (Objective 1; page 192 S; page 218 S/T)
2. a (Objective 2; page 194 S; page 220 S/T; video program)
3. c (Objective 2; page 195 S; page 221 S/T; video program)
4. b (Objective 3; pages 196–197 S; pages 222–223 S/T; video program)
5. a (Objective 5; pages 224–225 S; pages 258–259 S/T)
6. c (Objective 6; pages 224–225 S; pages 258–259 S/T; video program)
7. b (Objective 7; page 230 S; page 264 S/T)
8. b (Objective 8; page 232 S; page 266 S/T; video program)

Lesson 9

Matching

I.

1. a
2. f
3. e
4. d
5. j
6. c
7. g
8. i
9. b
10. h

II.

1. g
2. a
3. e
4. d
5. f
6. b
7. c

Completion

1. nucleotide, DNA, DNA, base pairs, transcription, polymerases, RNA, adenine, uracil, translation, polypeptide

2. triplets, codons, 64, amino acid, codons, amino acids, mRNA, rRNA, mRNA, tRNA, anticodons, [mRNA or codons], translation, tRNA, mRNA

3. mutations, mutations, mutagens, base pair, protein, sickle-cell anemia, protein, negative, repressors, inhibit, positive, promoters, speed up, tumors, malignant, cancer

Self-Test

(S after page numbers refers to the Starr textbook, S/T after page numbers refers to the Starr/Taggart textbook)

1. d   (Objective 1; page 202 S; page 230 S/T; video program)
2. d   (Objective 2; page 202 S; page 230 S/T; video program)
3. c   (Objective 2; pages 202–203 S; pages 230–231 S/T; video program)
4. b   (Objective 2; page 204 S; page 232 S/T; video program)
5. a   (Objective 3; page 214 S; pages 244–245 S/T; video program)
6. d   (Objective 3; page 214 S; pages 244–245 S/T; video program)
7. a   (Objective 4; page 216 S; pages 246–247 S/T; video program)
8. c   (Objective 5; pages 218–219; page 252 S/T ; video program)
9. a   (Objective 6; page 217 S; pages 250–251 S/T; video program)
10. b (Objective 6; video program)

Lesson 10

Matching

I.

1. d	6. g	11. f
2. n	7. i	12. h
3. o	8. a	13. e
4. b	9. k	14. l
5. m	10. c	15. j

II.

1. f	5. d	9. b
2. e	6. g	10. k
3. j	7. i	11. h
4. a	8. c	

Completion

1. fossils, biogeography, comparative morphology, gradual model of speciation, punctuation model of speciation, adaptive zone, adaptive radiation, extinct, background extinction, mass extinction

2. microevolution, allele, mutation, natural selection, gene flow, genetic drift, inbred, bottlenecked, genetic equilibrium

3. interbreed, species, species, allelic, mutation, gene flow, genetic drift, natural selection, genetic divergence, speciation

Self-Test

(S after page numbers refers to the Starr textbook, S/T after page numbers refers to the Starr/Taggart textbook)

1. c (Objective 1; page 244 S; page 276 S/T; video program)
2. b (Objective 2; pages 246–247 S; page 282–283 S/T)
3. a (Objective 3; page 246 S; page 283 S/T)
4. d (Objective 3; pages 246–247 S; page 283 S/T)
5. d (Objective 4; page 249 S; page 285 S/T)
6. b (Objective 4; pages 256–257 S; pages 292–293 S/T; video program)
7. a (Objective 5; pages 250–251; page 286 S/T)
8. a (Objective 6; pages 264–265; page 302 S/T)
9. c (Objective 7; pages 250–251; pages 286–287 S/T; video program)

Lesson 11

Matching

I.

1. c	5. e	9. a
2. i	6. f	10. h
3. l	7. k	11. d
4. g	8. j	12. b

II.

1. e	5. j	8. i
2. h	6. g	9. f
3. c	7. b	10. d
4. a		

Completion

1. indirect, Earth, other planets, organic molecules, amino acids, sugars, lipid protein membranes, computer simulations, organic molecules, enzymes, RNA

2. prokaryotic, archaebacteria, eubacterium, eukaryotic, cyclic, non-cyclic, oxygen, oxygen, aerobic, anaerobic, aerobic, mitochondria, chloroplasts, oxygen, ultraviolet radiation

3. comparative morphology, comparative biochemistry, macroevolution, fossil, anatomical, embryonic, nucleic acid, binomial, Linnaeus, phylogenetic, Monera, Protista, Fungi, Plantae, Animalia

Self-Test

(S after page numbers refers to the Starr textbook, S/T after page numbers refers to the Starr/Taggart textbook)

1. d (Objective 1; pages 274–275, 296–307 S; pages 312–313, 338–349 S/T; video program)
2. c (Objective 1; page 280 S; page 320 S/T)
3. c (Objective 2; page 292 S; page 334 S/T)
4. b (Objective 2; page 293 S; pages 334–335 S/T)
5. d (Objective 3; video program)
6. a (Objective 4; video program)
7. b (Objective 4; page 301 S; page 343 S/T)
8. a (Objective 5; video program)
9. b (Objective 6; page 305 S; page 347 S/T)
10. b (Objective 6; pages 296–297 S; page 338 S/T)

Lesson 12

Matching

I.

1. i	5. e	9. b
2. a	6. k	10. c
3. h	7. g	11. f
4. j	8. d	12. l

II.

1. d	4. c	6. b
2. a	5. g	7. e
3. f		

Completion

1. binomial, Linnaeus, genus, phylogeny, five, kingdoms, bacteria, protozoa, sporozoans, molds
2. prokaryotic, cocci, DNA, ribosomes, flagellum, binary fission, chemoheterotrophic
3. nonliving, noncellular, DNA, RNA, protein, lipid, protein, carbohydrate, DNA, host, DNA, proteins, are released from

Self-Test

(S after page numbers refers to the Starr textbook, S/T after page numbers refers to the Starr/Taggart textbook)

1. c (Objective 1; page 283 S; pages 328–329 S/T)
2. b (Objective 1; page 282 S; pages 322–323 S/T)
3. c (Objective 2; pages 282–283 S; page 328 S/T)
4. a (Objective 2; page 317 S; pages 362–363 S/T)
5. d (Objective 3; page 318 S; page 364 S/T)
6. c (Objective 3; page 317 S; page 362 S/T)
7. b (Objective 4; pages 312–313 S; page 355 S/T)
8. d (Objective 4; page 315 S; page 358 S/T)
9. a (Objective 5; pages 320–322 S; page 360 S/T)
10. c (Objective 6; pages 298–299 S; pages 340–341 S/T)

Lesson 13

Matching

I.

1. m	5. c	8. f	11. k
2. a	6. d	9. l	12. j
3. e	7. i	10. b	13. h
4. g			

II.

1. e	5. h	8. g	11. c
2. f	6. b	9. m	12. a
3. l	7. i	10. k	13. j
4. d			

III.

1. h	4. e	6. i	8. g
2. d	5. f	7. c	9. b
3. a			

Completion

1. algae, plants, fungi, root systems, vascular tissues, water conservation, reproduction, nonvascular, seedless, vascular, vascular, seeds, flowers

2. heterotrophs, decomposers, parasites, saprobes, spores, sac fungi, club fungi, zygomycetes, imperfect fungi

3. sponges, body plans, morphological features, behaviors, tissues, organs, mollusks, annelids, arthropods, echinoderms, chordates, chordates, vertebrates, endoskeleton, paired fins, respiratory structures

Self-Test

(S after page numbers refers to the Starr textbook, S/T after page numbers refers to the Starr/Taggart textbook)

1. d (Objective 1; page 334 S; pages 388–389 S/T)
2. a (Objective 1; page 336 S; pages 390–393 S/T)
3. d (Objective 2; page 342 S; page 400 S/T)
4. b (Objective 3; page 342 S; page 400 S/T)
5. a (Objective 4; page 344 S; pages 400, 402–403 S/T)
6. c (Objective 5; pages 350–351 S; pages 407–413 S/T)
7. d (Objective 6; page 358 S; page 418 S/T)
8. b (Objective 7; pages 358–359 S; page 419 S/T)
9. a (Objective 8; page 367 S; page 430 S/T)
10. b (Objective 9; page 372 S; page 436 S/T)

11. d (Objective 10; page 386 S; page 450 S/T)
12. b (Objective 10; page 390 S; page 454 S/T)

Lesson 14

Matching

1. i	9. k	16. q
2. l	10. b	17. m
3. u	11. d	18. g
4. e	12. v	19. p
5. r	13. j	20. o
6. f	14. c	21. n
7. a	15. t	22. h
8. s		

Completion

1. simple tissues, complex tissues, simple tissues, ground tissue systems, photosynthesis, complex tissues, xylem, phloem, dermal, xylem, dead, water, nutrients, phloem, alive, sugars, photosynthesis

2. leaves, roots, vascular bundles, leaves, stomata, electrical cohesion, cuticle, translocated, leaves, roots

3. apical, roots, shoots, apical, ground, primary, secondary, vascular, cork, stems, roots, woody, annual

Self-Test

(S after page numbers refers to the Starr textbook, S/T after page numbers refers to the Starr/Taggart textbook)

1. d (Objective 1; page 408 S; pages 482–483 S/T; video program)
2. c (Objective 2; page 408 S; pages 482–483 S/T)
3. a (Objective 3; page 409 S; page 483 S/T; video program)
4. b (Objective 4; pages 412–415 S; pages 484, 488–489 S/T; video program)
5. c (Objective 5; pages 418–419 S; pages 496–497 S/T; video program)
6. c (Objective 6; page 425 S; page 500 S/T)
7. b (Objective 7; video program)
8. c (Objective 8; page 430 S; pages 508–509 S/T)
9. a (Objective 9; page 432 S; pages 510–511 S/T)
10. d (Objective 10; video program)

Lesson 15

Matching

I.

1. b
2. c
3. g
4. j
5. h
6. d
7. a
8. e
9. k
10. i
11. f

II.

1. f
2. l
3. a
4. j
5. o
6. b
7. e
8. m
9. h
10. c
11. p
12. k
13. d
14. g
15. i
16. n

Completion

1. sexually, sporophyte, flowers, gametophytes, gametophytes, pollen grains, gametophytes, ovary, stigma, fertilization, zygote, embryo, embryo, endosperm, cotyledons, ovule, ovary, seed, fruit

2. asexually, parthenogenesis, shoot systems, runners, roots, shoots, shoots, buds, tubers, stems, induced propagation

3. auxins, gibberellins, cytokinins, dormancy, ethylene, rate, direction, tropic, light, gravity, temperature, photoperiodic, biological clock

Self-Test

(S after page numbers refers to the Starr textbook, S/T after page numbers refers to the Starr/Taggart textbook)

1. d (Objective 3; page 439 S; page 516 S/T)
2. b (Objective 3; page 439 S; page 516 S/T)
3. c (Objective 4; page 440 S; page 518 S/T)
4. d (Objective 5; page 442 S; pages 520–521 S/T)
5. a (Objective 6; page 446 S; page 530 S/T)
6. c (Objective 7; pages 448–449 S; pages 532–533 S/T)
7. d (Objective 8; page 454 S; pages 538–539 S/T)
8. a (Objective 9; pages 444–445 S; pages 524–525 S/T)

Answer Key 361

Lesson 16

Matching

I.

1. e
2. i
3. a
4. h
5. j
6. f
7. c
8. g
9. d
10. b

II.

1. c
2. b
3. f
4. h
5. d
6. e
7. a
8. g

III.

1. e
2. g
3. h
4. b
5. f
6. c
7. d
8. a

Completion

1. cells, tissues, organs, organs, organ system, internal environment, homeostasis, feedback mechanisms

2. epithelial tissue, connective tissue, muscle tissue, nervous tissue (any order); epithelial tissue; integument; exocrine, endocrine (any order); nervous tissue; neurons; electrical impulses; muscle tissue; contract; lengthen; connective tissue; bone, cartilage, blood, adipose tissue (any order)

3. integumentary system, skeletal system, muscular system (any order); integumentary system; epithelial cells; epidermis; connective tissue; dermis; bones; joints; fibrous tissue; cartilage, ligaments (any order); protection, support, (any order); mineral ions; tendons; myofibrils; sarcomeres; cross-bridge formation; myosin, actin (any order); sarcomere; sliding filament model

Self-Test

(S after page numbers refers to the Starr textbook, S/T after page numbers refers to the Starr/Taggart textbook)

1. b   (Objective 1; page 526 S; page 628 S/T)
2. a   (Objective 2; pages 462–463 S; pages 548–549 S/T)
3. a   (Objective 3; page 468 S; page 554 S/T; video program)
4. d   (Objective 4; pages 466, 526 S; page 629 S/T; video program)
5. c   (Objective 5; pages 463, 530 S; pages 634–635 S/T)

6. a   (Objective 6; page 531 S; pages 636–637 S/T)
7. d   (Objective 7; pages 528–529 S; pages 632–633 S/T)
8. a   (Objective 8; pages 464, 534 S; page 640 S/T)
9. c   (Objective 9; page 534 S; pages 640–641 S/T; video program)
10. b (Objective 10; page 537 S; page 644 S/T)

Lesson 17

Matching

I.

1. h	5. a	8. e
2. f	6. i	9. d
3. b	7. g	10. c
4. j		

II.

1. e	4. f	7. a
2. g	5. b	8. d
3. h	6. c	9. i

III.

1. i	4. b	7. f
2. h	5. d	8. g
3. e	6. c	9. a

Completion

1. arteries, capillaries, veins, blood, red blood cells, white blood cells, platelets, plasma, heart, ventricle, pulmonary circuit, atrium, ventricle, systemic, interstitial fluid, atrium

2. lymphatic system, lymph nodes, interstitial fluid, water, proteins, capillary beds, fats, pathogens, lymph nodes

3. erythrocytes, leukocytes, platelets, erythrocyte, anemia, sickle-cell anemia, leukemia, polycythemia

Self-Test

(S after page numbers refers to the Starr textbook, S/T after page numbers refers to the Starr/Taggart textbook)

1. c (Objective 1; pages 548–549 S; pages 656–657 S/T; video program)
2. d (Objective 2; page 542 S; page 650 S/T)
3. a (Objective 3; page 545; page 653 S/T)
4. c (Objective 4; page 548; page 656 S/T; video program)
5. b (Objective 5; page 551; page 659 S/T; video program)
6. d (Objective 6; page 554; page 662 S/T)
7. a (Objective 7; page 553; page 661 S/T; video program)
8. b (Objective 8; page 556; page 664 S/T)
9. c (Objective 9; pages 558–559; pages 666–667 S/T)
10. d (Objective 10; page 557; page 665 S/T; video program)

Lesson 18

Matching

I.

1. c
2. f
3. i
4. a
5. g
6. b
7. e
8. h
9. d
10. j

II.

1. d
2. a
3. f
4. c
5. h
6. j
7. b
8. e
9. g
10. i

III.

1. g
2. i
3. a
4. c
5. f
6. b
7. d
8. j
9. h
10. e

Completion

1. infection, toxic, neutrophils, eosinophils, mast cells, vasodilation, macrophages, complement, clotting, inflammation, swelling, contain, healing
2. four, chains, Y shape, light, constant, receptor, antigen, B cells, V, J, DNA, divide, clone
3. allergies, inflammation, mucous, genetic, environmental, allergens, pollens, IgE, mast, histamines, anaphylactic, antihistamines, desensitization

Self-Test

(S after page numbers refers to the Starr textbook, S/T after page numbers refers to the Starr/Taggart textbook)

1. c (Objective 1; page 564 S; page 672 S/T)
2. d (Objective 2; page 566 S; page 674 S/T; video program)
3. b (Objective 3; page 568 S; page 676 S/T; video program)
4. a (Objective 4; pages 568–569 S; page 676 S/T; video program)
5. a (Objective 5; pages 572–573 S; pages 680–681 S/T; video program)
6. d (Objective 6; page 570 S; page 678 S/T; video program)
7. a (Objective 7; pages 575 S; page 683 S/T; video program)
8. d (Objective 8; page 576 S; page 684 S/T; video program)
9. c (Objective 9; page 577 S; page 685 S/T; video program)
10. d (Objective 10; page 573 S; page 681 S/T; video program)

Lesson 19

Matching

I.

1. h
2. j
3. d
4. f
5. l
6. g
7. a
8. b
9. i
10. k
11. e
12. c

II.

1. f
2. e
3. c
4. b
5. g
6. d
7. h
8. a
9. i
10. j

III.

1. b
2. d
3. g
4. e
5. c
6. a
7. h
8. j
9. f
10. i

Completion

1. oxygen, carbon dioxide, exchange, surface-to-volume, diffusion, ventilation, gills, transport pigments, red, binds, tissues, circulatory, flow, blood

2. obtain, gas, dispose, integument, skin, gills, tubes, spiders, lungs, energy, environment, air sacs, respiratory, amphibians, countercurrent, fish, alveoli

3. environmental pollution, cigarette smoking, bacteria, bronchioles, bronchitis, emphysema, marijuana, petroleum, fungi, chlorine gas, breath, cancer

Self-Test

(S after page numbers refers to the Starr textbook, S/T after page numbers refers to the Starr/Taggart textbook)

1. b (Objective 1; page 584 S; page 692 S/T)
2. d (Objective 1; video program)
3. b (Objective 2; page 584 S; page 692 S/T)
4. c (Objective 3; page 587 S; page 695 S/T; video program)
5. d (Objective 4; page 590 S; page 698 S/T)
6. a (Objective 5; page 588 S; pages 696–697 S/T; video program)
7. c (Objective 6; pages 592–593 S; pages 700–701 S/T; video program)
8. a (Objective 7; pages 592–593 S; pages 700–701 S/T)

9. d (Objective 8; pages 594–595 S; pages 702–703 S/T)
10. d (Objective 9; video program)

Lesson 20

Matching

I.

1. f	5. k	9. j
2. h	6. b	10. e
3. c	7. i	11. d
4. l	8. a	12. g

II.

1. j	6. k	10. b
2. d	7. m	11. f
3. e	8. c	12. a
4. i	9. h	13. g
5. l		

III.

1. f	5. g	8. j
2. c	6. i	9. d
3. b	7. h	10. e
4. a		

IV.

1. h	5. j	8. f
2. e	6. c	9. b
3. d	7. i	10. g
4. a		

Completion

1. homeostasis, internal environment, nutrients, waste materials, carbohydrates, proteins, fats, vitamins, minerals, nutrients, villi, small intestine, bloodstream, fats, lymph system, adipose tissue, ammonia, urea, uric acid, water, excretion, evaporation, respiration

2. digestive systems, oral cavity, saliva, polysaccharides, saliva, esophagus, stomach, gastric fluid, protein, small intestine, liver, gallbladder, pancreas, digested, absorbed, colon, excreted, rectum

3. extracellular fluid, urinary system, kidneys, nephrons, glomerulus, water, small solutes, filtered, Bowman's capsule, capillaries, tubules, secreted, urinary tract, urine, ureter, urinary bladder, urethra

Self-Test

(S after page numbers refers to the Starr textbook, S/T after page numbers refers to the Starr/Taggart textbook)

1. b (Objective 1; page 600 S; page 712 S/T)
2. d (Objective 2; pages 606–607 S; pages 716–719 S/T; video program)
3. d (Objective 3; page 604 S; pages 715–717 S/T)
4. b (Objective 4; page 601 S; page 713 S/T; video program)
5. c (Objective 5; page 610 S; page 722 S/T)
6. c (Objective 6; page 620 S; page 732 S/T; video program)
7. a (Objective 7; pages 621–623 S; page 733 S/T; video program)
8. a (Objective 8; page 621 S; page 735 S/T; video program)
9. d (Objective 9; page 623 S; page 735 S/T)
10. d (Objective 10; page 625 S; page 737 S/T)

Lesson 21

Matching

I.

1. i	5. j	9. g
2. c	6. a	10. d
3. l	7. f	11. e
4. b	8. k	12. h

II.

1. j	5. a	9. k
2. l	6. i	10. g
3. h	7. b	11. f
4. d	8. c	12. e

III.

1. a	5. e	8. c
2. g	6. d	9. b
3. i	7. h	10. f
4. j		

Completion

1. peripheral nervous system, neurons, nerves, somatic, autonomic, sympathetic, parasympathetic, central nervous system, spinal cord, brain, brain

2. chemical, synapse, dendrites, cell body, axon endings, sodium ions, resting membrane potential, action potential, action potentials, neurotransmitters, plasma membrane, action potentials, sodium-potassium pumps

3. sensory neurons, how many, how often, somatic, movement, position, pressure, tissue damage, pain, heat, cold, hearing, vision

Self-Test

(S after page numbers refers to the Starr textbook, S/T after page numbers refers to the Starr/Taggart textbook)

1. c (Objective 1; page 474 S; pages 559, 560 S/T; video program)
2. a (Objective 2; page 474 S; page 560 S/T; video program)
3. c (Objective 3; page 478 S; pages 564–565 S/T; video program)
4. b (Objective 4; page 481 S; pages 566–567 S/T; video program)
5. c (Objective 5; page 488 S; page 578 S/T; video program)

6. d   (Objective 6; page 502 S; page 590 S/T)
7. c   (Objective 7; page 496 S; pages 590–591 S/T)
8. a   (Objective 8; pages 498–499 S; pages 596–597 S/T)
9. b   (Objective 9; pages 500–501 S; pages 600–601 S/T)
10. c   (Objective 10; pages 490–491 S; pages 584–585 S/T; video program)

Lesson 22

Matching

I.

1. c
2. f
3. a
4. j
5. e
6. d
7. b
8. h
9. i
10. g

II.

1. g
2. b
3. c
4. f
5. a
6. h
7. j
8. k and d
9. e
10. i

III.

1. h
2. j
3. a
4. e
5. i
6. d
7. g
8. b
9. c
10. f

Completion

1. gland, exocrine, endocrine, islets, hormone, delta, alpha, glucose, beta, insulin, muscle, glycogen, decreases, glucagon, increase, somatostatin

2. pituitary, hypothalamus, oxytocin, posterior, interstitial, anterior, produces, releasers, inhibitors, eight, interaction

3. feedback, TSH, thyroid, iodine, homeostasis, dwarfism, somatotrophin, gigantism, acromegaly, pineal, increases, sleepy, decreases, light, biorhythms

Self-Test

(S after page numbers refers to the Starr textbook, S/T after page numbers refers to the Starr/Taggart textbook)

1. b (Objective 1; page 461 S; page 547 S/T)
2. c (Objective 2; page 508 S; page 610 S/T)
3. a (Objective 3; page 508 S; page 610 S/T)
4. b (Objective 4; page 512 S; page 614 S/T)
5. a (Objective 5; page 516 S; page 618 S/T)
6. d (Objective 6; pages 515, 518–519 S; pages 617, 620–621 S/T)
7. c (Objective 7; pages 514, 520 S; page 616 S/T)
8. d (Objective 8; pages 510–511 S; pages 612–613 S/T)

9. a (Objective 8; page 511 S; page 613 S/T)
10. d (Objective 9; page 510 S; page 612 S/T; video program)

Lesson 23

Matching

I.

1. e	6. n	11. m
2. l	7. d	12. c
3. h	8. g	13. k
4. j	9. a	14. f
5. b	10. i	

II.

1. f	5. h	9. a
2. c	6. b	10. e
3. i	7. k	11. j
4. d	8. l	12. g

Completion

1. sperm, testes, eggs, ovaries, sperm, LH, FSH, pituitary, FSH, LH, estrogen, progesterone, eggs, endometrium, ovary, endometrium, LH, ovulation, corpus luteum, progesterone, estrogen, implantation

2. oocytes, sperm, fertilization, zygote, cleavage, embryo, amnion, chorion, allantois, yolk sac, endoderm, mesoderm, ectoderm, differentiation

3. in vitro fertilization, abstinence, vasectomy, tubal ligation, diaphragms, condoms, pills, teratogens, placenta, tobacco, alcohol, drugs

Self-Test

(S after page numbers refers to the Starr textbook, S/T after page numbers refers to the Starr/Taggart textbook)

1. d (Objective 1; pages 632–633 S; page 746 S/T; video program)
2. d (Objective 2; page 639 S; page 753 S/T; video program)
3. a (Objective 3; pages 642–643 S; page 766 S/T; video program)
4. c (Objective 3; page 647 S; pages 768–769 S/T)
5. a (Objective 4; pages 648–649 S; pages 768–771 S/T; video program)
6. b (Objective 5; pages 656–658 S; pages 778–779 S/T; video program)
7. d (Objective 6; page 661 S; pages 758, 783 S/T)
8. d (Objective 7; pages 662–663 S; pages 784–785 S/T)
9. b (Objective 8; page 659 S; pages 780–781 S/T)
10. d (Objective 9; video program)

Lesson 24

Matching

I.

1. g
2. b
3. e
4. j
5. c
6. a
7. f
8. h
9. d
10. i

II.

1. f
2. c
3. h
4. d
5. a
6. e
7. g
8. b

III.

1. b
2. f
3. d
4. a
5. e
6. h
7. c
8. g

Completion

1. clumped, age, exponential, reproductive, rN, J-shaped, food, living space, carrying capacity, logistical, $r_{max} N [(K - N)/K]$, S-shaped, density dependent, carrying capacity, density independent, drought, volcanic eruptions

2. demographic transition, preindustrial, low, transition, increase, industrial, zero population, postindustrial, Brazil, France

3. neutral, mutualism, parasitism, commensalism, symbiosis, life, predation, interspecific competition, regulate, niche, community

Self-Test

(S after page numbers refers to the Starr textbook, S/T after page numbers refers to the Starr/Taggart textbook)

1. d (Objective 1; video program)
2. b (Objective 2; page 676 S; page 798 S/T; video program)
3. c (Objective 3; pages 676–677 S; page 798 S/T; video program)
4. d (Objective 4; page 677 S; page 799 S/T; video program)
5. c (Objective 5; page 685 S; page 809 S/T; video program)
6. b (Objective 6; page 684 S; page 808 S/T)
7. c (Objective 7; page 683 S; pages 806–807 S/T)
8. a (Objective 8; page 690 S; page 814 S/T)

9. d   (Objective 9; page 692 S; pages 816–817 S/T)
10. a (Objective 10; page 698 S; page 824 S/T)

Lesson 25

Matching

I.

1. d
2. a
3. g
4. c
5. j
6. b
7. f
8. i
9. e
10. h

II.

1. h
2. b
3. d
4. j
5. g
6. a
7. c
8. i
9. e
10. f

III.

1. b
2. f
3. i
4. a
5. j
6. e
7. c
8. h
9. d
10. g

Completion

1. disturbances, energy, nutrients, ecosystem, autotrophs, sun, heterotrophs, recycle, consumers, decomposers, detritivores, inputs, computer, predictive, model

2. biogeochemical, fixation, nitrogen, plants, animals, fungi, nitrification, nitrite, nitrate, denitrification, fertilizers, fossil

3. atmosphere, warm, east, Peru, Humboldt, evaporation, coasts, precipitation, loss, productivity, drought, global, populations

Self-Test

(S after page numbers refers to the Starr textbook, S/T after page numbers refers to the Starr/Taggart textbook)

1. c (Objective 1; page 708 S; pages 838–839 S/T)
2. c (Objective 2; pages 708–709 S; pages 836–837 S/T)
3. b (Objective 2; pages 710–711 S; pages 838–839 S/T)
4. c (Objective 3; page 713 S; page 841 S/T)
5. a (Objective 3; page 713 S; page 841 S/T)
6. a (Objective 4; page 721 S; page 851 S/T)
7. d (Objective 5; pages 726–727 S; pages 856–857 S/T)
8. d (Objective 6; pages 731–732 S; pages 860–862 S/T)

9. b (Objective 7; pages 740–741 S; pages 870–871 S/T)
10. d (Objective 8; pages 708, 721 S; pages 836, 851 S/T; video program)

Lesson 26

Matching

I.

1. c
2. f
3. d
4. h
5. j
6. b
7. a
8. e
9. g
10. i

II.

1. g
2. e
3. i
4. a
5. c
6. j
7. h
8. b
9. f
10. d

III.

1. e
2. a
3. c
4. f
5. j
6. b
7. g
8. d
9. i
10. h

Completion

1. cropland, more grazing, fuel, lumber, temperate, Brazil, Columbia, erosion, weather, biodiversity, flooding, extinctions, greenhouse, grasslands

2. climatic, solar, photovoltaic, electricity, solar-hydrogen, emissions, turbines, batteries, biomass, geothermal, hydroelectric, fusion, unlimited, implosion, laser

3. biosphere, desalination, freshwater, wastewater, primary, chlorination, secondary, microbial, tertiary, technology, pollutants, expensive, quality

Self-Test

(S after page numbers refers to the Starr textbook, S/T after page numbers refers to the Starr/Taggart textbook)

1. b (Objective 1; pages 752, 764 S; page 894 S/T)
2. b (Objective 2; pages 766–767 S; pages 896–897 S/T)
3. c (Objective 3; page 754 S; page 884 S/T)
4. b (Objective 3; pages 754–755 S; pages 884–885 S/T)
5. d (Objective 4; pages 762–763 S; pages 890–891 S/T)
6. a (Objective 5; pages 757–761 S; pages 888–895 S/T)
7. d (Objective 6; pages 752–753 S; pages 882–883 S/T)
8. c (Objective 7; page 766 S; page 896 S/T)
9. d (Objective 8; video program)
10. c (Objective 8; page 767 S; page 897 S/T)